Aesthetics, Necropolitics, and Environmental Struggle

Critical Art Ensemble

Other Critical Art Ensemble Titles from Autonomedia

The Electronic Disturbance

Electronic Civil Disobedience and Other Unpopular Ideas

Flesh Machine
Cyborgs, Designer Babies, and New Eugenic Consciousness

Digital Resistance
Explorations in Tactical Media

The Molecular Invasion

Marching Plague
Germ Warfare and Global Public Health

Aesthetics, Necropolitics, and Environmental Struggle

Critical Art Ensemble

Autonomedia

Acknowledgments

CAE would like to thank the following people for the illuminating thoughts and discussions that contributed to the completion of this project: Jim Fleming, Lewanne Jones, Ala Plástica, Brian Holmes, Eduardo Molinari, Azul Blaseotto, Graciela Carnevale, Rebecca Schneider, Nicola Triscott, YoHa, Eugene Thacker, Nell Tenhaaf, Paige Sarlin, Marc Bölen, Trevor Paglen, Andreas Broeckmann, Diana McCarty, Konrad Becker, Amy Balkin, Alan Braddock, Natalie Jeremijenko, Tania Bruguera, Natasha Lushetich, Brooke Singer, Sarah Lewison, Derek Curry, Jennifer Gradecki, and all our friends at Parco Arte Vivente.

We send special thanks to Tony Conrad (1940–2016) and Tim Rollins (1955–2017), who were inspirations to CAE from the start. We will always be grateful for their decades of support.

Autonomedia
PO Box 568 Williamsburgh Station
Brooklyn, New York 11211-0568 USA
Phone: 718-963-2603
www.autonomedia.org

Contents

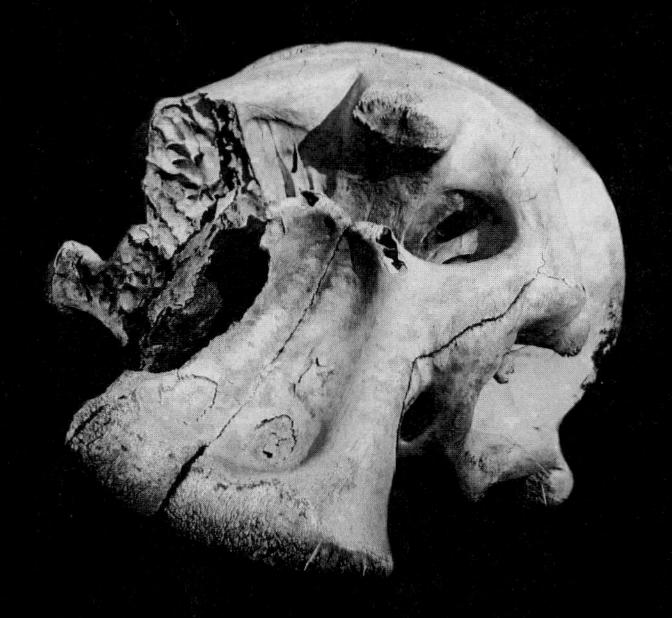

Elephant

Preface

The Indecipherable and the Indeterminate

Is global capitalism indecipherable? CAE would not blanket a field so large and complex with a single tendency, but we do think it is an ascending one as system complexity and process velocity increase. When complexity grows faster than intelligence can adapt and velocity moves beyond human ability to process and respond, the fluctuating and impermanent condition of the indecipherable (re)emerges. Currently, it is unfortunately functioning as a root cause in all the various crises of capitalism, including the crises in the military, financial, and environmental orders.

This tendency was almost immediately visible in the military order as soon as the United States and its coalition partners attacked, devastated, and destabilized Iraq. The neoconservatives' misguided ideological belief that the US military and company could conquer and maintain order while a new Western-leaning pseudo-democratic (puppet) state was birthed can only be described as a milestone in political stupidity. What is happening instead is a kaleidoscope of violence with so many moving parts, so many illusions taken to be real, and so many shifting terrains that the situation has become pragmatically unintelligible. This is not to say that histories of the crisis cannot be constructed. In hindsight a reasonable narrative can be spun pointing to the problems arising from continued Western military aggression, the reemergence of tensions stemming from tribal, ethnic, and religious differences suppressed and/or reorganized by colonization, along with problems derived from de-Baathification and fundamentalist zealotry. Be that as it may, no one in a position of power is able to turn what knowledge there is into an operational plan to resolve the conflicts. The current choices in terms of response to the problems of the Middle East are all bad: more military interventions doomed to failure, peace talks that never produce peace, or simply allowing death to rule the region.

US presidential policy itself is indicative of the inability to decipher the situation. Consequently, a plausible plan of action cannot be constructed. For the Bush administration, the Rumsfeld military doctrine of "if you can't solve a problem, make it bigger" was a roll of the dice that was at best a long shot. The doctrine holds that if a problem can be expanded, more and more power vectors will be pulled into the conflict until a critical mass of political and military will arises that shall, by whatever means, bring the conflict to a halt. Conservatives championing accelerationism is odd, but clearly can happen. The Obama policy answered this with the opposite strategy of "Don't do stupid shit." Which is to say, do what is best for the US, do not go all in, and extract ourselves from the conflict to the fullest extent possible. We have no idea how to solve the problem, but at least we will not be the ones who make the conflict worse, and fewer assets will be wasted. But the Goldilocks method of deploying just the right amount of violence is difficult when a situation is not quantifiable. Some on the left argue that if presidential policy was simply free of imperialist intentions, the situation would be improved. Possibly, but the residue of imperialism is too ingrained in the territory to stop the violence. In the Middle East today, even if the West had a policy of immediate withdrawal and nonintervention, it would not stop the violence among those jockeying for power in the destabilized region, which in turn pushes the refugee crisis ever further.

This leads us to the next point of indecipherability: Is it still possible for neoliberal powers to construct a military that makes sense within the logic of the US and Western military order itself, or functions with some semblance of efficiency and usefulness in the time of hegemonic global capitalism? CAE will leave it at "no one knows." Imagining the kind of weaponry, training, and deployment that will be suitable for future conflicts in which the actors are not nation-states, while simultaneously keeping military production aligned with entrenched economic and political interests in the homeland appears to be nearly impossible. The ecology of future conflict is yet to be understood, leading to a serious lack of predictability, which, as in this situation, intensifies problems in all the spheres of crisis. For the military, it means a failure to reorganize its social ecology and the continuation of near-useless Cold War military production.

Let us consider another subject that has had quite a bit of attention recently: contemporary finance, and to be more specific, the crash of 2008—an event that can only be understood in hindsight, partly because of the tremendous amount of obfuscation built into the process, partly because of corruption, and partly because of its relation to the indecipherable. Once some transparency entered the process, understanding the criminality driving the crash was fairly simple. The financial instruments that developed around mortgages, in conjunction with complete corruption in the ratings agencies, allowed for a game that paired the appearance of maximum risk with maximum certainty for those in the know. The bad paper generated in this exchange could only end with blind major investors (pensions funds, unions, cities, etc.) and subprime mortgage holders taking a mighty haircut. Financial institutions imploded that were undercapitalized and overextended with toxic holdings or were insuring toxic holdings. Where CAE is skeptical is the implication among conspiratorialists that various CEOs and other business leaders willingly gambled their banks or investment houses because they were confident that the government would bail them out, no one would be criminally charged, and their personal fortunes would remain untouched. And in terms of prestige, there has to be some cognitive dissonance among those who lost or crashed their own institutions.

This is what brings us to the indecipherable. The institutional implosions were partially due to the fact that captains of finance were blindsided because their explanatory models were completely outdated. Most notably, the extremely popular Black-Scholes option pricing formula from 1973 was in no way equipped for the twenty-first-century market. It was particularly weak on predicting volatility. Prior to the crash, the Chicago Board Options Exchange Volatility Index (VIX) was typically moving between 10 and 25. In 2007, the VIX was at 11, and it averaged 19 from 1990 to September 2008. In October 2008, after the collapse of Lehman, it reached an intraday high of 89.5. According to the model, such a shift should be impossible. In conjunction, the S&P moved 10 percent (1 percent is the norm) in a single day ten times over a six-month period. No one anticipated this, nor had the means to anticipate it. No market model could predict it. Predictability had reached a limit.

When we look at Enron being rated at AA- four days before default, or the numerous AAA-rated ("impossible to fail") instruments that failed in the 2008 crash, we know that corruption and criminality played a big part. But CAE does not think it is an exhaustive explanation. We do not think any ratings agency, bank, or investment house had any clue as to how much damage they were actually doing, and what the real consequences would be (especially to themselves). To place this all in the category of precision conspiracy is giving the financial class too much credit.

On a smaller scale, one can look at the two primary "flash crashes" of May 6, 2010, in the US and October 2013 in Singapore. No one knew what happened in any technical sense, only in a general sense, with the most popular theory being that high-velocity black-box trading had reached a point of excess and needed limits. These occurrences are not so interesting for the losses posted (in the case of the US crash, losses were modest), but as examples of the kind of indecipherable accident that will occur as more processes are removed from human agency and turned over to autonomous machines with massive computational abilities—the cause of the accidents is lost in the data.

In the financial and military orders, the crises of the elite differ from those of everyone else. Those who reap little benefit from capitalism know that our crisis concerns how to change the profound economic inequality inherent in the system, as opposed to the elite crisis of stabilizing a wild and unpredictable system of exchange. And, whereas the capitalist crisis in the military primarily involves navigating the complex military field in order to create armed forces that are more efficient and better aligned with the current moment in history, our crisis is imagining how to stop imperial use of military power around the globe. When turning to the next area of crisis, the environment, the indecipherable does emerge again, but there are two significant differences. First, whether people support capitalism or not, we share with global capitalists a common problem of complexity related to planet/environmental maintenance. Everyone can agree that the earth must remain a planet that can sustain human life (even if for some capitalists this is believed to be a short-term concern), and that complexity is making it difficult to find solutions.

The second significant difference is that while the military and financial orders evolved in complexity alongside knowledge systems, the environment is simply a given that struck us in its vast entirety from day one. Humans have the motivation and the methods to analyze and understand their environment, but unfortunately the job is too large. There are not enough biologists, ecologists, climatologists, and geologists to even construct more than the smallest of ecological maps, nor is there enough computational power for the job. This is particularly true in the discipline of ecological studies. All the living creatures of the world are not close to being catalogued. We do not know the bodies and the behaviors of a numerically overwhelming number of flora and fauna. Reams of undiscovered species are on the planet, and many more are only vaguely understood. Since the most basic knowledge of all of the forms of life that inhabit the earth is far from complete, how can we expect to map all of the interrelations, including those involving nonliving elements of the environment? Even relatively small subsystems are difficult or impossible to analyze and understand in their totality. At this point the problems begin to parallel what we have seen in the last two examples—predictability is lost as complexity increases. Strategies aimed at resolving environmental problems become impossible. Humans know we are in the Anthropocene, but have only the most tentative and incomplete idea of what this means.

When the problem is simple (human-initiated climate change is bad if a rich and diverse ecosystem is of value), consensus is quick to emerge among experts and those who interpret the world using reason. But putting aside the voices of greed and self-interest that retard ecological action, how can scientists, activists, and the concerned move forward from the point of defining the culprit as human production of greenhouse gases, and explaining the consequences of inaction? Since we have no ecological, economic, or sociological model for this crisis, predictability is lost to innumerable competing hypotheses on what the best strategy or even what a merely helpful tactic might be. Massive amounts of trial-and-error hypothesis testing is in front of us, in which we do not know what the consequences of any particular action will be. In the age of the Anthropocene, the culture/nature system has become a singular system of interrelations and interdependencies. Achieving any predictive power becomes difficult with so many variables from all parts of nature and society at work—there is no model that takes both of these once-separate systems into account as a singular system. As we shall see, this is a problem (clashes of interests and needs among stakeholders) even with such relatively tiny environmental and ecological questions as how large the deer population should be on public land in the Northeastern US. Who knows what will occur when actions affect production and life on a global basis? CAE hopes to show that the failure to slow trajectories that we know are fatal to humans and most life on earth is aggravated further by the lack of explicit necropolitics.

Knowing that inaction will certainly end badly, activists and concerned citizens seem to fall back on "saving" whatever we can with whatever resources available. Save the rain forest from almost every variety of exploitation humankind can generate, save the whales from Japanese whalers, save the community gardens from developers and speculators, save the national parks and wilderness areas from the extraction industries, and so on. Not that this is unhelpful, but we all silently know that it is not the answer we are looking for, in that the climate crisis and its consequences keep on accelerating.

Activists and concerned human inhabitants of earth endlessly find ourselves back in the realms of affect and aesthetics, which is to say that the choice of actions is ultimately arbitrary. CAE does not mean this in a negative way. As we have stated, one thing we do know is that doing nothing will go badly, which means we must gamble with acting, even if only through trial and error, with the lives of humans and nonhumans alike on the line. Having done experimental work in cultural activism for the past thirty years, we are very familiar with this method. We do not know what the outcome of a particular project will be (which is what makes it experimental in the cultural sense), and we do not know how it will contribute to the general political and cultural tendency we are hoping to participate in by doing the action or project. The outcome is indeterminate (for a greater explanation of politicized cultural indeterminacy, see Appendix III). Let us say that we are doing a project that we hope will participate in a general tendency toward environmental justice. We know at the outset that there is no consensus as to what ecological justice is, nor do we know how to achieve it. We crudely understand it tactically—for example, this action stopped a wilderness area from being fracked. But we do not know how to get to the ultimate goal in this case—for the extraction industry to start contracting instead of expanding. We can save a tiger, but that is not slowing the acceleration of extinction.

CAE is often asked, "What good are these cultural actions? When has a work of art changed anything?" CAE has to admit it is true that a work of art has rarely changed anything. However, framing the question in this way is the curse of individualism, wherein somehow a single person, or a single action, is supposed to change the world. That is not how change happens. The *aggregate* of cultural action *over time* is what has changed many elements of society. Positive change could come when all the gamblers are acting, experimenting, and conversing, because emerging through all the actions and exchanges come possibilities, and as these possibilities emerge, they mingle and recombine themselves into different arrangements that may eventually reveal how to understand, act, organize, and achieve environmental justice, sustainable environmental practices, and biodiversity. This model has been the model of many cultural activists since the early 1990s, and it is certainly the one CAE has promoted since *The Electronic Disturbance* (1994). And this model for engaging our relation to indeterminacy is one we share with tens if not hundreds of thousands of authors. As in the other aforementioned struggles, it appears to CAE that in the struggle for environmental justice and biodiversity, this is the best means (if not the only one) to cope with our baseline ignorance and the indeterminacy of our actions. This method requires some very unpleasant thinking as well—thinking about ecology through death rather than life. Thinking anthropocentrically instead of internaturally. Thinking of this universe as absent of compassion other than what is generated by humans. Thinking of each of these subjects through the frame of struggle, and how that may lead us to the bets we make. In this book, we are trying to get out of the box of prevailing wisdom, and hence the conceptual punching and flailing we are herein engaging.

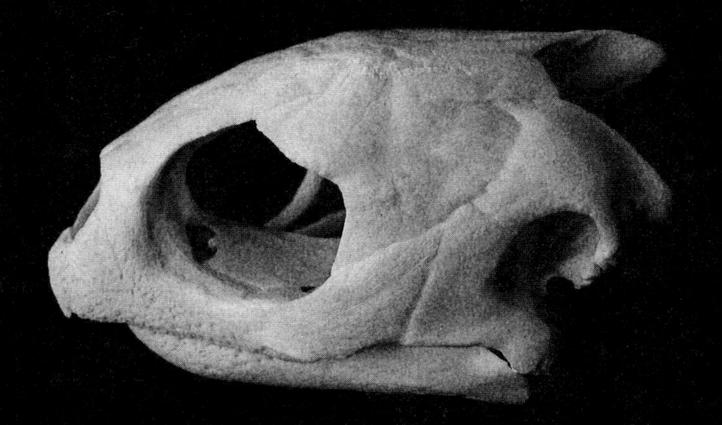

Loggerhead Sea Turtle

Introduction

The Demonic Effect of a Fully Developed Idea

Over the past twenty years, a central point of exploration for CAE has been revolutions and crises related to the environment, including the operational impact of molecular biology, climate change, struggles against the extraction industries, species extinction and rescue, environmental injustice, conservation strategies, internaturalism, and so on. As long as the intervention or project made tactical sense, we would move forward, guided as much by affect as we were by reason—generally a very potent cocktail of human possibility. This method of action appeared to us as the standard for activists and people concerned with this great variety of environmental issues. As environmentalism fractured into a huge spectrum of specialized causes, the discourse and the actions among concerned, more specialized populations were also divided both philosophically and technically (also usually a good thing for a many-faceted resistance trying to establish itself and attempting to find its way in very difficult terrain). Be that as it may, certain questions, facts, and principles keep haunting us, as people work for environmental amelioration from the narrow perspective of their own primary issue or set of related issues. Among the problems that has bothered us most is the absence of explicit necropolitics from strategic and tactical plans that clearly have an implicit, often unspoken, necropolitical dimension. For example, take the *Deep Ecology Platform* (1984), written by two luminaries of the environmental movement, Arne Naess and George Sessions, as a condensation of fifteen years of environmentalist thinking:

1. *The well-being and flourishing of human and nonhuman life on Earth have value in themselves (synonyms: inherent worth, intrinsic value, inherent value). These values are independent of the usefulness of the nonhuman world for human purposes.*
2. *Richness and diversity of life forms contribute to the realization of these values and are also values in themselves.*
3. *Humans have no right to reduce this richness and diversity except to satisfy vital needs.*
4. *Present human interference with the nonhuman world is excessive, and the situation is rapidly worsening.*
5. *The flourishing of human life and cultures is compatible with a substantial decrease of the human population. The flourishing of nonhuman life requires such a decrease.*
6. *Policies must therefore be changed. The changes in policies affect basic economic, technological, and ideological structures. The resulting state of affairs will be deeply different from the present.*
7. *The ideological change is mainly that of appreciating life quality (dwelling in situations of inherent worth) rather than adhering to an increasingly higher standard of living. There will be a profound awareness of the difference between big and great.*
8. *Those who subscribe to the foregoing points have an obligation directly or indirectly to participate in the attempt to implement the necessary changes.*

CAE would think that anyone with a heart would enjoy the poetic elegance of these statements, and sympathize with the affect they induce—but do they make sense? For us the text is problematic. We know that humans, in any concentrated form, have never been good for a healthy environment for humans or for most nonhuman creatures (with the exception of our microbiome). Not since pre-agrarian social formations, and with a total of 250 million humans or fewer on the planet, have we, by accident of circumstance, not harmed the environment, which makes CAE a little worried when we read principle five.

CAE also has to ask what is meant by "well-being and flourishing." Other than associating it with utility (which lends itself poorly to poetry), it appears that the reader is invited to project any meaning that suits their interests or desires. The same could be said for "vital needs," which apparently give humans some necropolitical rights in regard to the nonhuman. The fantasy engine of consciousness is invited to rev up and complete this open-ended text.

The next phrase, the declaration that human and nonhuman life have "value in themselves," is also a problem. Values are hierarchical conceptual constructions produced by humans and shared by humans (or imposed upon them). They are not objects waiting to be discovered.

"Diversity" is the next over-aestheticized word within this manifesto. Diversity in *social* and *cultural* formations is certainly a characteristic to be desired, as it enriches human experience and opens new possibilities and potentials in those who participate in these diverse formations. But do we need an ecosystem that maximizes diversity? The perspective necessary for the answer to be yes is an anthropocentric one (just as with culture). If the world were once again populated by nothing more than single-celled creatures, neither they nor the earth nor the universe would care one bit. Diversity in the natural world appears to CAE as an aesthetic and/or economic value. Since only humans care for economy and aesthetics, negative views of anthropocentrism (a spillover from science) need to be reconsidered. In other words, there is no *scientific* reason why biodiversity is valuable, as such judgments are beyond the limits of science.

But soon we get to what this platform is really about: a substantial decrease in the human population, which is presented as an ideological adjustment rather than the massive material cataclysm that such an imperative would entail! How do you eliminate billions of people from the planet? Who has to kill their germ line, and who gets to reproduce? Which cultures are eliminated? All of them? Does accepting these principles necessitate participation in eco-nihilism? Various forms of eco-nihilism such as that suggested here are becoming givens among radical environmentalists. The one element this drastic proposal does bring to light is that producing life and producing death—biopolitics and necropolitics—are indivisible opposites. The real challenge lies in balancing the critique of the two. Currently, discourse is skewed toward biopolitics in both its negative and positive forms, while necropolitics remains rather neglected. After all, promoting life is so much easier than taking a stand on what should die, and how death should be operationalized. Moreover, once necropolitics is more than just acknowledged, but made visible, the company in which we find ourselves is quite unpleasant. Moving into active planning within the parameters of a zero-sum game puts a person at the table with Malthusians, Spencerians, colonialists, fascists, corporatists, developers of total war, and most worrisome of all in its unexpectedness, a variety of everyday bureaucrats, technocrats, and policymakers just doing their jobs. If nothing else, CAE has to admire the courage of Naess and Sessions in saying that humans are the problem, and that we need to rid ourselves of as many as possible. Unfortunately, they do not have the courage to explain how we should do this. Other radical greens do, as we shall see—but they leave the tactical choices up to the revolutionaries involved.

For CAE as well, we do not have a plan on how to activate necropolitics in a manner that makes sense. We are stuck outside the ramparts, unable to proceed any further, and until a language(s) is developed to acknowledge necropolitics and develop policy through democratic means (although environmentalism and democracy may be incompatible), the population of environmentally conscious actors (we hesitate to call it a movement) will only be able to continue to "save" life guided by aesthetic choice in a fragmented, tactical manner (which in so many cases is only a deferral rather than a saving). As a result, the real environmental crises will remain unaddressed. Obviously, CAE does not have an answer here either, and many of the criticisms, contradictions, and concerns contained in this book are aimed at ourselves as much as anyone else. We hope to name some of the conceptual demons hiding within us, so we may at least get to the point of exorcising those that paralyze us and cause as much internal discomfort as the external environmental degradation we are witnessing.

North American Bison

1

Necropolitics and the Dark Comedy of the Posthuman

Twenty years ago, CAE wrote its first essay on the posthuman ("Posthuman Development in the Age of Pancapitalism"), and now we are back grappling with this topic once again. The upside is that very little has changed over the past two decades. The posthuman fantasy is still just that, but the fact that this fantasy remains so entrenched in the collective imagination of technocrats, engineers, solutionists, and digital enthusiasts speaks to the power of nihilistic desire in humans—a desire for the end of humanity. This desire is what truly distinguishes those who champion the posthuman. The posthumanists are not pretenders or reformers. They have no interest in panhumanism like the so-called postmodern antihumanists (a very poor choice of words), nor are they promoting some newly revised philosophy of humanism. They are revolutionaries dedicated to creating an explosive movement in evolution in which a new creature specializing in intelligence is formed, and/or to eliminating humans from much or all of the earth.

Humans

Unfortunately, for the purposes of this essay, a model of what a human is is necessary, so that it might function as a point of contrast to the posthuman.

What CAE is about to offer is meant neither as a universal nor as a philosophically robust definition. We are only offering an imperfect, operational model that consists of eight points of tremendous elasticity. We are speaking of tendencies that, when bundled in various configurations, could represent many human variations.

1. Flesh. Humans are tied to the organic.
2. Consciousness. Humans are not only conscious of objects in the world, but are also conscious of themselves as objects.
3. Cognition. Humans are generally beyond stimulus/response, and have the ability to think and even, at times, reason in the gap between the two.
4. Language. Humans can manipulate signs and symbols, and through this manipulation can communicate with one another in complex ways, which allows for the manufacture and accumulation of knowledge.
5. Sociality. As Aristotle profoundly stated, humans are social animals.
6. Technical proficiency. Humans can extend their bodies and minds through the use of technical objects.
7. Mortality. Humans die, and are aware they will die.
8. Biological reproduction. Humans have the ability to reproduce their population.

These tendencies can be placed into concrete situations, and in those situations different assemblies of the human can be made. The situation could be very simple, such as person in a deep coma that may just be living flesh, or it could be a very complex interrelationship that would be more typical of life experience.

Cyborgs

One distinguishing characteristic of humans is their ability to radically enhance their capabilities and manipulate their environments through the use of technical objects. Consequently, humans who are at the more utopian and complex end of the technosphere are forever flirting with the idea of their technologies becoming integrated with their bodies and brains to an extent of complete interdependence. Conversely, those at the more apocalyptic end of the spectrum have a dread and fear of being fully integrated with their machines and becoming slaves to them and/or to the masters of the machines. Humans who have the power to do so appear to be more than willing to walk up to the line, but generally do not cross it. In the "happier" places within the complex technosphere, we see endless populations of digital zombies armed with phones, pads, and laptops, and permanently at work—either intentionally and directly working in the virtual marketplace in order to survive, or working unintentionally by doing "recreational" personal and social data entry for corporations and security agencies. Fortunately for them, they can still unplug at the end of the day. Those caught in the manufacturing and service hells of the global order such as call centers, data entry centers, and digital sweatshops also remain attached to their machines, but are recognized less as humans and more as necessary parts of a greater machine.

Full integration of human and machine to a point of complete interdependence seems to be the pivotal point here. Preparations are of course being made. Soldiers are now weapons systems, and under combat conditions probably relieved by the prospect. On the civilian side of the digital empire, as indicated above, classes of technocrats are filled with anxiety at the prospect of being without their devices. Some humans have entered a time when even sleep management needs a device. Data must always be produced.

For the past twenty years, CAE has heard the constant argument that to have a pacemaker is a life-and-death integration of flesh and technology, so cyborgs already exist. If we use a flexible definition of a cyborg, perhaps. What holds CAE back from endorsing this position is that a cyborg as described here is not posthuman—in fact, it helps to explain why humans have not taken the next evolutionary step into full cyborgian existence. If we examine the popular fictional cyborgs that CAE would call posthuman (DC Comics's Cyborg, Darth Vader, and the like), not only was the technology fully integrated with its organic platform, it enhanced the power of the individual to move far beyond human limits. A person with a pacemaker is only being returned to a human range of physical health and normality. CAE does emphasize the word "human" in this case. The narratives guiding the fictional examples of the posthuman above center around the urge to either "improve" human abilities (move faster, elevate intelligence, become indestructable) and use those improvements to better the world (Cyborg is a member of the Justice League), or to enhance oneself in order to release a contained inner evil (or, to be less metaphysical, a maliciousness that will yield power in the form of domination). Both narratives are quite intriguing, and clearly so given the global fascination with superheroes and supervillains who surpass the boundaries of human norms. The question then becomes: Why is research not being poured into transplantable technologies and the techniques of organic/synthetic integration that will propel humans into the posthuman? Why isn't the market demanding these products of science fiction at once?

While humans can figuratively, and at times literally, love their devices and long for full integration, they also have millennia of purity myth constantly telling them that recombination is either evil or bad. This myth is so pervasive that it makes its way into many laws and norms, and severely retards any evolution toward cyborg existence. In this myth in its many, many forms, the creation of life is not in the domain of humans. It is either in God's domain or in Nature's. When humans interfere in this process, bad things happen, and the person or persons who do the interference are generally severely punished. Whether it is Icarus trying to fly or Dr. Frankenstein trying to create a human, the consequences are disastrous. Most of the monsters who populate stories and myths are recombinant creatures (which may in fact be the definition of the monstrous). Vampires, werewolves, demons, and devils are ungodly combinations of the Creator's work, or a deep perversion of Nature. In turn, this piece of ideology has been used to maintain all manner of secular purity. Social and economic separations and hierarchies are reinforced and maintained though the appeal to purity and through the fear of the recombinant.

The mixing of human and machine, the organic and the inorganic, violates these categories of separation, and with this comes very deep skepticism about the morality of cyborg existence beyond that of normal human function. Two quick examples: First, one of the bulwarks stopping cyborg evolution is that medicine sees intervention into the human body for any purpose other than to improve the health of the individual as frivolous and/or unethical/immoral. CAE has to wonder whether the argument used for cosmetic surgery (that it improves the mental health and quality of life of the patient) could apply to cyborg surgery. At the moment, apparently, no. Risking the dangers of medical intervention to satisfy a desire unproven to help a patient is not going to occur. Moreover, the doctor's insurance company will reinforce this idea, as will the patient's insurance provider.

Second example: In 2010, the artist Wafaa Bilal had a titanium plate implanted in the back of his head, to which a camera was attached that constantly took photos. CAE will let Bilal explain his reasons. What is important to this discussion is the reaction. Some of Bilal's students at New York University and the university itself objected, and he had to cover the camera while on campus. Why did this camera have to go? The argument from the university was that the camera had to be covered up or turned off for privacy reasons. Such an argument is patently absurd in an age of ubiquitous surveillance. For any resident of New York City, or for anyone who carries a cell phone for that matter, surveillance is simply a fact of everyday life. Not only is the university, as well as NYC, awash in cameras monitoring the vast majority of public space, but every student and faculty member is carrying one or more cameras at all times, and it is more than likely that these devices are tracking and gathering other data on each individual throughout the day. Something else must be at stake. The problem was not the act of surveillance; rather, it was the *form* of surveillance. CAE would argue that the camera's embeddedness in flesh was too unnerving. It constituted a form of surveillance that was too far beyond the norm by calling attention not just to itself, but to a future flesh-and-technology integration that appeared perverse. We might say a similar thing about the failure of Google Glass. Even that simple level of flesh/machine integration for the purpose of superior performance was too unnerving.

It will be very difficult to change these norms. This is why the cyborgian posthuman is not emerging. Corporations in the business of recombinant life (particularly those in the food supply industry) are finding this out as well. These corporations have no problem with the purity myth for purposes of social exploitation, but need to build in an exception so people will perceive their products with something other than suspicion and disgust. Cyborgs share this problem. We are no closer to this posthuman form than we were five decades ago. At present, cyborgs are nothing more than a compelling fiction.

Transhumanists and Extropians

Happily, this thought fad among the STEM (science, technology, engineering, and math) professions remains in the previous century, although some harmful residue still exists. That it is of any interest at all is due to the fact that it does present a variety of ideas about how to end humanity and transcend into a new kind of being with fewer "limitations." Pundits such as Max More, Ray Kurzweil, Aubrey de Grey, and Hans Moravec believe that this transition could happen in their lifetimes. Of course, such thinking could only occur in a bunker so isolated from the rest of the world and the daily life of most humans that boundaries between unbridled speculation and material reality could be ignored—a place where science fiction could be taken as reasoned reflection on the near future and mixed with the wild optimism of rapid progress. This loosely knit alliance of ideas regarding the advancement of technology in relation to the advancement and ending of humans is an odd form of right-wing accelerationism.[1] The faster STEM professionals can end humans, the more interesting the world will become. It is not that these believers are hoping for the worst; they are not. They are just done with being human, and find the Faustian bargain to be a smart way to relieve their boredom and frustration over species limitations.

What also seems to help their optimism is their belief in capitalism. The question "Are you attracted to innovative, market-oriented solutions to social problems?" posed by More to help interested individuals decide if they might be an Extropian is telling. The market is the means to solve social problems (it does not make them), and it will act as an extra line of defense against the deployment of new technology that could be construed as antihuman. Unfortunately for us in the twenty-first-century present, solutionism (market-driven technological fixes for all our problems) has not gone away. That element of the transhuman gospel still remains a fixture in the ideology of STEM.

Where might technological development lead us? Humans could become cyborgs (as Moravec believes), or we could upload consciousness and live within the wires and circuits in the disembodiment of pure thought. And if we want to return to our former human selves, we can preserve our bodies (or at least some cells), and technocrats will recall us when there is a cure for whatever damaged us. We can live the best of both worlds. If disembodiment does not sound attractive, then immortality might be a more appropriate selection. There are two ways this could be achieved. One is through nanotechnology, as popularized by Eric Drexler in *Engines of Creation* (1986), in which nanobots programmed to maintain, or even morph, our bodies keep us healthy and our identities flowing. The second path to immortality is through biological intervention, of which there are two primary possibilities. This may occur, on the one hand, as genomic manipulation that turns humans into recombinant creatures of our own design. (This is one reason why morphological freedom and reproductive freedom have been defended by transhumanists as civil liberties.) The other biological possibility, usually associated with de Grey, is that biologists will find a way to circumvent that which causes humans bodily damage unto death. Quite rightly, the processes that kill humans are known: mutations in chromosomes, mutations in mitochondria, junk inside and outside of cells that the body cannot eliminate, cell loss, extracellular protein cross-links, and cellular senescence. De Grey is convinced that therapies to address these problems will be created in his lifetime, and they will allow him a long enough life span that he will see these therapies perfected, enabling him to become immortal. Strangely enough, this hope has made its way into the twenty-first century.

The technocratic solution to necropolitics is one of transformation—an avoidance in which humans can at least pretend to be beyond death. This refusal, combined with a belief that the market and accelerated technological development will also provide protection from human misdeeds and public policy errors, allows questions over the environment to be avoided. STEM will take care of it. The world will be a better place the sooner it becomes a fully engineered and managed environment. CAE would prefer that madcap ideas about nanobots and uploading were all that survived of transhumanism, but unfortunately it is the transhumanist philosophy of solutionism that has continued to sustain its relevance. This is a position that lacks any regard for necropolitics in that it refuses to confront human finitude as a fundamental reality, or to admit that mass extinction continues unabated and that the human species is among the contemporary contenders for this possibility of extermination.

At the other end of the political spectrum, we can find a posthuman future that also thrives on acceleration. This vision rests on sustaining the current environment without humans or with limited human presence in limited areas. As with the transhumanists, there is no language for necropolitics, so while mass death is often implied in this discourse, it is never directly addressed.

Green Posthumans

In the introduction, we met the first of the posthuman deep greens and their poetic musings on saving the environment. They set the tone for the belief that the human experiment has been an unfortunate development for the natural world. Humans need to drastically slow, if not halt, their reengineering of the planet and bring about a vast reduction of their numbers. As we shall see with later deep green radicals, this applies to attendant species as well. Domestic animals and plants, those that have thrown their evolutionary lot in with humans and are thus a part, however unwittingly, of the current "biotic cleansing," are deserving of the same fate of rapid reduction if not elimination. Yet in all the present-day environmental discourse, death is never directly addressed. It is always only implied, and left to be the elephant standing in the room. We can take as examples the moderate to center-right environmentalist and biologist Edward O. Wilson and radical green thinker and activist Lierre Keith. CAE has tremendous respect for the work of both to the extent of considering them proven friends of environmental struggle. We are looking at their work here only to illustrate a problem that plagues the entire movement across its political continuum, and not as an individualized accusatory complaint. CAE is asking what an environmentalist necropolitics would look like. Because until one explicitly emerges, the movement is doomed, if for no other reason than an inability to even discuss achievable, strategic objectives.

In 2016 Wilson published *Half-Earth: Our Planet's Fight for Life*. In it, he paints a picture of what is at stake in current environmental struggle by indicating what humans will lose and what we have already lost in regard to biodiversity. The book is a compelling figurative call to arms and describes the sense of alarm that people should feel regarding the current ecological crisis of mass extinction.

> The only hope for the species still living is a human effort commensurate with the magnitude of the problem. The ongoing mass extinction of species, and with it the extinction of genes and ecosystems, ranks with pandemics, world war, and climate change as among the deadliest threats that humanity has imposed on itself.

CAE could not agree more with this description. The problem is when Wilson gets to the solution, and the solution is hinted in the title of the book: half of the earth's lands and oceans should become a posthuman environment. One would think that the majority of the book would be an explanation of how this could be done without the employment of outright atrocity. That is not the case. Only a very small portion of the book is dedicated to how this will occur, and most of this portion is marked by the sort of exhaustingly Pollyannaish optimism that is generally reserved for the transhumanists.

Wilson argues that the population problem will solve itself, because future citizens will choose not to over-reproduce. They will reproduce only their own numbers or fewer, or not at all. The facts that he presents to promote this position seem to undermine it, in CAE's reading. Yes, there has been a slight deceleration in population growth. And yes, the US and Europe are at zero population growth. And yes, in countries where women have greater economic and political power, fertility has dropped. However, Wilson goes on to admit that population growth is not going to stop and that by the end of the century the population will be somewhere between 9.6 billion and 12.3 billion. The real problem with Wilson's solution, however, is that on a planet already strapped for resources, once half of them were off limits, the population rate would indeed go down, but not primarily because women would choose to reproduce less—starvation, malnutrition, inadequate healthcare, and conflict over limited resources would be more likely causes. Wilson also seems to think that most of the world's women will share in the power status of European and North American women by the end of the century, thus lowering population growth even further. CAE fails to see any evidence to justify this belief. Regardless, if the population problem is not solved, it does not seem that the extinction crisis will be either.

Wilson goes on to tell us that fewer resources will be needed because the free market guided by high technology will accelerate a shift from extensive economic growth to intensive economic growth. We will have an economy of quality over quantity in which twentieth-century conspicuous mass consumption will be shunned. Extrapolating from the text, the Wilsonian future will be in dense cities where people stay in their luxury pods and connect, consume, do business, and so on, using ever-improving high technology. Fewer resources will be used and less pollution generated. GM crops will improve crop yields for abundant food. Apparently, humans will still go outside, as Wilson mentions in an offhand remark that anyone will be able to visit any reserve. This is odd, since reserves require human intervention such as management and policing, meaning that half the earth would *not* be left in a posthuman condition.

Here we find ourselves back in league with the transhumanists. Solutionism has made its way into neoliberal environmental discourse, resulting in the great faith that the free market and technological fixes will save the planet, humans, and the diversity of the natural world. Leaving aside the fact that the world described above sounds closer to a prison than a utopia to CAE, what of all the problems that technology and the free market cannot solve? Sure, they can give us faster computers at lower prices for better shopping, but can they stop natural disasters? Can they eliminate water shortages? Can they keep free-market greed from destroying economies and facilitating depressions? Can they stop military adventurism? They have failed to do anything of the kind so far. Moreover, the assumption that the free market works best for all would seem to conflict with historical events. In fact, what the global free market has brought is a misery for the majority so vast that the philosopher of misery himself, Arthur Schopenhauer, would be shocked.

This kind of whitewashing of such a difficult problem is to be expected. After all, Wilson is writing a semi-popular book, the audience for which is not interested in chapters on land use law and administration or resource management and conservation. At the same time, part of the reason for *Half Earth's* poverty of language is the human tendency to pretend that necropolitics does not exist, short of ethnic cleansing or genocide. For example, US congressional committees are considering repealing the Affordable Care Act. Should they succeed, the result will be near or immediate death for tens of thousands and shorter life spans for those who are denied medical coverage (which will be in the millions). Congressional committees are essentially deciding who will live (wealthy people) and who will die (poor people), and doing so in a legitimized manner, as they were seemingly fairly elected to make such decisions, much to the surprise of a benighted electorate. The current necropolitical system, as reflected in this example of the bureaucratic order of death, is a Malthusian form in which excess populations are designated by implication and left to die through malevolent neglect. This is the form of necropolitics in which we are all implicated but can pretend that we are not. One reason the Malthusian form of necropolitics continues ever onward is because it is the easiest to ignore. Unfortunately for many Americans, necropolitics has come out of its dark world, and is demanding attention.

The radical answer to reform, education, and market solutions that is latent in extremist environmental discourse is that civilization must be destroyed. In the 2013 conference anthology *Earth at Risk: Building a Resistance Movement to Save the Planet*, Lierre Keith lays out a plan and gives her literal call to arms (Keith's and other conference lectures are also available on YouTube). Keith appears certain about two principles. First, that any complex division of labor from agrarian society forward is devastating to the environment, and hence unacceptable. (She gives agriculture a necropolitical name: "biotic cleansing.") Second, the current system of global neoliberalism will hit a crisis point and implode. Given these principles, she takes an accelerationist position, and argues that eco-warriors everywhere should do whatever they can to hurry this implosion, which most significantly includes an asymmetrical war against civilization. The endgame of this activity is to eliminate humans from large portions of the planet, and reduce their numbers and capabilities so that humanity is returned to a sustainable hunter-gatherer form of community. The goals, in her own words, are:

> Part 1: To disrupt and dismantle industrial civilization; to thereby remove the ability of the rich to steal from the poor and the powerful to destroy the planet.
>
> Part 2: To defend and rebuild just, sustainable, and autonomous human communities; and as part of that, to assist in the recovery of the land.

With regard to the first goal, Keith explains:

> Finally, we aim for coordinated, multiple attacks using surprise; what we are ultimately after is cascading systems failure.
>
> The point is not to make a statement. The point is to make a decisive material impact. In other words: we bring it down.

Keith advocates this type of military adventurism with true passion and seriousness. Her example of how such a plan is already being operationalized is the Movement for Emancipation of the Niger Delta (MEND). From 2006 to 2009, this native resistance movement attempted to expel Royal Dutch Shell, Chevron, and Exxon from the delta region, and see the oil wealth divided among the region's people (despite an amnesty deal with the government, there continue to be flare-ups). However, even if they are successful it is very difficult to see how the area could be recuperated, because it is so badly polluted that the ecosystem has crashed. These are experienced fighters who understand guerrilla tactics and strategy, and have managed to make their enemies pay in blood and assets for their continued occupation of the region. They also have popular support in the area, which contributes to the success of their operations. Using this model of fighting, Keith hopes to attack high-value targets that will lead to maximum damage to the system. She is not thinking, like Earth Liberation Front, of attacking logging camps or universities. She wants to sabotage sites like power stations and water systems. Hers is a plan of major destruction. The plan seems to be to end civilization through mass sabotage. She does not explicitly advocate killing people directly, although CAE does not see how that could be avoided given the targets that interest her. As CAE reads her, Keith believes that her movement would be popular enough to get a critical mass of above-ground support in the way of aid and good will, but at the same time knows that this will be a tiny portion of the population, and unlike with MEND, the overwhelming majority of the population will be against the activity.

Admittedly, CAE has never heard of something like this before in a secular context. Keith's plan is a milestone in necropolitics. It is biblical, an Antichrist level of death and destruction. To get to the magic hunter-gatherer number of less than 250 million, around seven billion people need to die. This makes Wilson's half-earth strategy look modest in its death toll (although both are equally unlikely, which is to say not in the realm of possibility). CAE cannot even begin to imagine the kind of fanatical commitment to the aesthetic of biodiversity that could lead a person to stand before an audience and advocate creating an apocalypse that would kill almost every person now on the planet. As if this doctrine could not get more unfathomable, two other unusual points are thrown into the plan. First, Keith claims to wants to facilitate this final solution out of love; and second, she is a utopian who believes that once the dust clears, the survivors, if any, will live in hunter-gatherer peace and harmony organized around a progressive identity politics and environmental stewardship. Given the current political climate of antiterrorism and the sophistication of contemporary surveillance and tracking, CAE cannot see a campaign like this lasting even as long as the Symbionese Liberation Army, Red Army Faction, or Black Liberation Army. And we would not be the least bit surprised to see, if there were to be any planned mission, a preemptive strike by security forces. After all, if one is going to be a leader in the destruction of civilization, it is probably best not to talk about it in public forums and publish books and videos about it.

Ideas on the posthuman are remarkable only in the level of fantasy they engage. Clearly they provide excellent fodder for science fiction, where they can provide hours of amusement, but as human strategy to solve problems in the world they are far less useful. What would be helpful, as CAE hopes this analysis shows, is an emergent necropolitics—a popular means to speak about death in an organized manner and in relation to policy. Such activities are happening among various bureaucrats and officials, and of course the militaries of the world, on a daily basis as they make decisions about who lives and dies, but no one else has legitimate claim to do so. For a common citizen to do so is to be at best callous and uncaring, and at worst monstrous. Unfortunately, full-spectrum democracy is not possible without it, and how it emerges if neglected is in fantasies like the ones we have been examining in this essay. CAE believes it is time to pull necropolitics out of the abyss of the inhuman.

Note

1. In the US, accelerationism has become quite common on the right, even though historically it is a concept associated with left-wing entities typically on the periphery of Marxism. As of late, the accelerationist creed has been accepted by the extreme right—the so-called alt.right and the neoconservatives. The alt.right notion of the deconstruction of the administrative state is a kamikaze mission to eliminate as many bureaucracies as can be eliminated and to push the rest into dysfunction. The sooner this happens, the sooner the next chapter of American history can be birthed. The neocons, on the other hand, used accelerationism as a strategy for mission management. This is what they did in Iraq—accelerated the crisis in the hope that as more and more states and actors were pulled into the conflict, and as more and more interests became compromised, something would have to give (as with left accelerationists, what will "give" is never stated because it is an unknown). In the case of the Iraq war, this unknown moment was ostensibly when meaningful negotiations or total war would happen, and a new Middle East would be created.

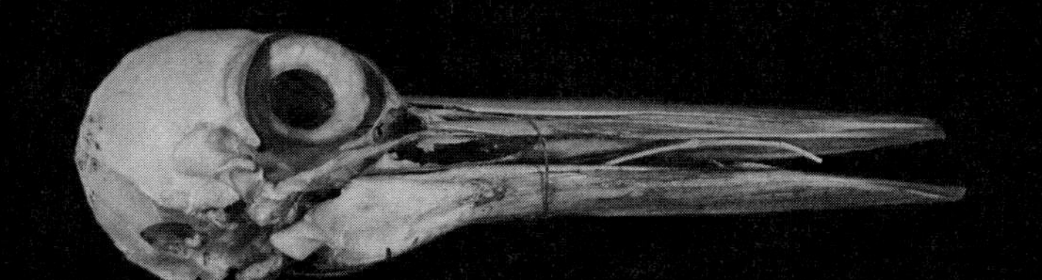

Ivory-Billed Woodpecker

2

Anthropocentrism Reconsidered

CAE will begin this chapter by telling a nonfiction story that is impossible to accurately tell. The reason it is impossible to tell is because we do not know the intention of the key actor. The intention must be inferred from the occurrence, and this process of interpretation is very much open to the biases and expectations of the narrator. Be that as it may, we will try to tell it.

This story is about a bonobo named Kanzi, who currently lives in the Iowa Primate Learning Sanctuary in Des Moines. Kanzi was born in captivity, and has an ape mother, Matata, and an adoptive human mother, biologist Sue Savage-Rumbaugh, who raised Kanzi from shortly after his birth. Sue immersed Kanzi in an English language environment in which he began to learn the meaning of words, which he expresses through the use of a lexigram keyboard (although he also has limited vocalizations and knows some sign language). Kanzi appears to know over four hundred words.

Here is the story as narrated by Bill Fields, a sanctuary researcher and one of Kanzi's internatural partners. Kanzi has a room that is separated from an attached room by glass, so that he can be observed without anyone having to enter his space. On this particular day, Kanzi's human mother was in an argument with a colleague over what video format should be used for the archive. As the argument grew increasingly heated, Kanzi banged on the glass, prompting Bill to enter his room. Kanzi "said" to Bill that he needed to punish the person who was acting aggressively toward Sue. Bill replied that he could not intervene in this way. He deferred to human etiquette. Kanzi said that Bill should punish the aggressor or he would bite Bill. Bill did not intervene. Twenty-four hours later, as Kanzi was being transferred from one to space to another, he broke away from Sue. He ran into Bill's office, where he bit off one of Bill's fingers and mangled his hand. Consequently, Bill refused to speak to Kanzi. Kanzi called for Bill repeatedly over a period of months. He wanted to resume their association. Bill insisted that he would not see Kanzi until Kanzi apologized. Kanzi replied that he had not done anything that required an apology. After eight months of Bill's silent treatment, Kanzi finally apologized with a hug and a submissive scream reinforced by saying "yes" when asked if his humble actions constituted an apology.

Beyond the material events (and short of using the most tortured of science-speak, even these are nearly impossible to describe without prejudice), how can we interpret what happened in this story? Was there a linguistic exchange? Was this a collision of bonobo and human culture in internatural space? Or, more specifically, was this a collision of bonobo and human models of justice in internatural space? Did Kanzi give Bill an ultimatum (a very complex thought pattern)? Why did Kanzi wait twenty-four hours to bite Bill, or could we say, carry out the threat? Or were the bite and the "ultimatum" unrelated? Why did Kanzi think an apology was not necessary, and why did he finally agree to apologize? For those charged with the imperative of never anthropomorphizing, this story has to blow some circuits. As nonrational as it may be, and certainly unscientific, the desire to go all in on anthropomorphizing this situation is amazingly intense. CAE wants to agree with Savage-Rumbaugh that Kanzi is an "ape at the brink of the human mind." We begin to wonder whether anthropomorphism, or its sibling anthropocentrism, are always such bad things. For understanding another species within strict scientific parameters, they are probably so, but in terms of connecting or bonding with other species or ecological systems less common to human experience, perhaps not. For building a more empathetic awareness of the natural world, and for all the internatural relationships we have on a daily basis, anthropomorphism and anthropocentrism might be necessities. There is an affirmational desire in anthropomorphism and anthropocentrism that can be harnessed in a manner that motivates people to work for the environment and nonhuman creatures alike.

If we examine the well-known animal-saving initiatives, anthropomorphism plays a significant role. The "cute factor" is often sited as a key nonrational motivator in these campaigns—but *why* is a creature cute or charismatic? CAE would say this is an anthropomorphic response. Baby mammals are cute because they are an extension of human babies, with their large, probing eyes and petite size. Moreover, we can project onto them qualities that we find desirable in our own children, such as innocence and playfulness. The animals with which humans strongly identify are those that are easiest to anthropomorphize—the majestic big cats, the soaring eagles, the playful dolphins, and the cuddly pandas. At other times the bond exists because we believe they are close to us in consciousness, as with whales and great apes. When we examine the continuum of our attraction and disgust for other animals, it seems very apparent that those animals most like us skew toward the attractive side and those that we identify with least skew toward the repulsive. For example, humans tend to like animals that are dry. We tend not to care for wet and slimy creatures. We tend to prefer fur to scales, and vertebrates to invertebrates. Good news for tigers and polar bears; bad news for the tumbling creek cave snail (which is also endangered).

Our anthropocentric feelings for these animals result not only from their perceived similarity to us, but also from our desire to be like them. We project the qualities we find noble onto nonhumans and let them mirror these qualities back. Associating human characteristics with animals is engrained in us, and often, in its positive form, manifests in objects like totems, mascots, coats of arms, and other insignia. Of course, through negative aesthetic bias we associate the worst with nonhuman life as well. No one wants to be a weasel, a slug, or scum, but for the most part the human inability to disassociate our own qualities from those of animals is the basis for empathy for them, for the environment that sustains them, and for the planet we share. Anthropocentrism and the humanism that emerges from it is also the basis of the ideology that has supported environmental progress in the US. CAE will now move away from this rather whimsical discussion to examine how the forces of humanism (with all its anthropocentrism) and true antihumanism (with all its contempt for any life beyond the ego) have battled over the environment.

Why Conservative Americans Love John Locke

CAE will just come right out and state that any discourse or movement (with few possible recent exceptions) in the US that has been of benefit to the environment has had anthropocentrism at its foundation, and conversely, that most of the bad that has happened has been in part due to its absence. To be sure, anthropocentrism can be very ugly given that its key principles are that humans are at the center of all life, and that the good is measured by what is beneficial to humans. But we should not take these principles in isolation. Once put into a social and historical context and framed within a humanistic perspective, these prideful notions may not be as environmentally dangerous as they first appear—particularly if we contrast them with those of capitalism in its raw neoliberal form. In US-style neoliberalism, the *individual* human is the center of all things and the good is measured by how much benefit an individual reaps. Humanity as such is not a meaningful category, and is certainly not worth any investment as that would lessen or negate potential benefits to the individual. In neoliberalism, it is "every man for himself," and all other subjects and objects are merely resources to be used to better one's station or to lower someone else's. Investment in environmental sustainability is a waste since the individual is finite. Resources need only last for a lifetime. But CAE is getting ahead of ourselves.

If we want to understand the environmental debate over land and resources in the US, we must review the work of John Locke (1632–1704). Locke was a British revolutionary and philosopher who championed the goal of destroying the monarchy and ending the divine right of kings. He wanted to reset the chain of being so that humans were at the top, sharing the world as equals, and creating forms of governance that reflected these new social relations. Lockean philosophy was of exceptional significance in colonial and revolutionary America, and would continue to be, long after many of its key ideas had been disassociated from Locke in the popular imagination and transformed into traditional political wisdom. Locke's ideas were pivotal in the American development of ideas about individual rights, land use, and property rights, and none of Locke's works was more significant than *The Second Treatise of Government* (1690).

For Locke, the starting place for understanding property and government was providence. Humans are on earth for a purpose, and that is to move through time improving their quality of life. God the Creator has made humans and earth as a perfect potential (divine property). Human purpose is to develop oneself and the environment as a way to serve God by completing His work, thus creating natural property. The key to development is labor. Labor is important on two levels. First, it is through labor that value and property are created. An apple on a tree is but a potential placed there by God to be used by people. It is incumbent upon an individual, when confronted by such potential, to pick the fruit, lest it go to waste. Should an individual not answer the call, thus allowing the fruit to rot, he (Locke acknowledged only male labor) would be guilty of idleness and refusing God's command to complete His work. Conversely, if the fruit is picked, through his labor an individual may now take ownership and may consume the fruit or take it to market to sell to others, thus bestowing upon the apple a value that it did not have prior to the act of picking it.

Second, labor becomes important in and of itself. In the North, among the Puritans and other Protestants, this made perfect sense. When a person works he is following in the footsteps of the Creator, and perhaps more importantly, shielding himself so that sin may not find its way into his soul. Idle hands are the Devil's workshop, so the best protection from temptation is constant work. For those who had broken from the Catholic tradition with its emphasis on prayer, reflection, and introspection as means to commune with that which is holy and as means to protect oneself from sin, the only safe engagement with a world in which worship was limited was work. This same ethic was popular in the Protestant South as well. Idleness was not acceptable. This concept was used as one of the justifications for slavery. Many slaveholders were of the belief that they were protecting the souls of their slaves by keeping them in constant toil, for otherwise they would assuredly be led into temptation. Poor white people were browbeaten with this principle (for surely they would not be poor if they were working as they should). A "cracker"—someone who is "cracked" in the head because they are ignoring their duty to God by living in idleness—was profoundly looked down upon (and often considered to be of a station lower than a slave for his refusal to work). A man's relation to hard labor was a direct measure of his character. As these notions evolved into partisan political ideology, they became the basis for the conservative contempt for any type of welfare, as well as for those who accept it. Conservatives did not recognize any type of structural economic, historical, or social disadvantage. Blame for an impoverished state of being always fell upon the individual as a personal problem (a character defect), and never on the system as a social problem.

There is one more very sad chapter regarding idlers in America, and that is the story of the indigenous North Americans. Lockean theory was taken to mean that those who built a lifestyle around the sin of idleness, mocking God with their refusal to answer the call of providence by developing the land, should be forcibly removed or exterminated so that those of a more industrious character could develop the resources. The indigenous people with their nomadic low-production communities were acting as a hindrance to God's work. Many of those participating in the destruction of native peoples and/or their cultures believed they were carrying out the will of God, and thus doing what was right for the environment and the nation. CAE should note, however, that like the doctrine of "manifest destiny," this application of Lockean philosophy was contested. Conquest by imperialist means was not a consensus position.

The consequence of this collection of ideas was profound in the American experience. The idea that one should work hard and develop the land in a frontier nation was a given. In terms of the colonizers, it was in everyone's best interest, from the wealthy to the poor. Locke was widely embraced from north to south, and most certainly among the founding fathers. The westward expansion was nonstop, always with the idea of public land development, and in turn led to the enclosure and privatization of massive tracts of land. However, Locke did provide some limits on development that were very similar to anarchist ideas on personal property. Locke believed that land enclosure should be limited to the amount of land an individual (or family, because the Bible recognized this social unit) could work. If land was lying fallow, or if its fruits were left rotting, another individual would have the right to claim that land or product. So not only were there limits to claims, but the claim had to be continually worked. There was no resting on past achievement.

A second important development for environmental relations is contained in this ideological package. Locke is suggesting—and the founding fathers and subsequent administrations, all the way through the nineteenth century, did declare—that all public land was open to development. And all land that was not in private hands was public. This was a practical way to build a nation, but as an ideology it became an environmental disaster as the land ran out. In all fairness, Locke could never have foreseen the development of forms of power beyond that of the flesh of people and animals. Steam and electrical power were too far on the horizon. Nor could he ever have dreamed that the US would make corporations individuals. Locke was not a supporter of mass inequality, so CAE believes he would have had to seriously update the limits he proposed had he been able to see what development had wrought by the late twentieth century, and how his system of limits had been inverted into a justification for limitless development and resource exploitation.

A final consideration for environmental relations is the meaning of "waste." This would become a very contested word by the end of the nineteenth century, but for the early history of the US, the meaning was fairly stable and enjoyed a near consensus. Waste was the failure to maximize the potential of a given resource. Locke thought of this in primarily agrarian terms. Ten acres of cultivated land could produce more than one hundred acres left to nature. Locke appears to associate high production with wealth and abundance, and not with any negatives, as God has provided all that humans could ever need. The implications of this idea of waste for the actual land can be somewhat surprising to the environmentally sensitive reader. "Wilderness" becomes an extreme negative in this ideological system because it is immediately associated with waste. Wilderness is any land not being developed to its full economic potential. Wilderness is the land equivalent of a human who is an idler. They are the abject and the unacceptable. Like waste, wilderness would also become a contested concept by the end of the nineteenth century.

The Lockean State

Locke begins his consideration of governance with a reflection on humans in a state of nature. This presocial (although prepolitical would be a better term) state is set in contrast to life in the Hobbesian world famously expressed as "nasty, brutish, and short." While Locke would agree with Hobbes that in a state of nature each individual is sovereign, he did not believe that the state of nature was anarchy. There were laws of nature that guided interrelations among individuals. There was a sense of justice and injustice. This tempered the population of individuals, so there was no war of all against all, but there were most certainly disputes, and disputes were settled by the individuals involved in the disagreements in both just and unjust ways. For Locke, this was not optimal, and thereby fell into the category of the undesirable, if not of sin, particularly when disputes ended in death. No means existed to justify death in a dispute, and murder was a violation of divine property. Humans were to be developed like any other resource, so murder, even in the event of perceived injustice, was unacceptable. On the other hand, Locke thought that a sentence of death from a fair and impartial judicial system as a way to remedy an injustice was acceptable, and he considered it to be a legitimate use of political power.

In order to avoid problems of injustice among sovereigns, associations of individuals came together to make a contract to form a government. Political power is thus brought into being, and is described by Locke thus:

> *Political power*, then, I take to be a *right* of making laws with penalties of death, and consequently all less penalties, for the regulating and preserving of property, and of employing the force of the community, in the execution of such laws, and in the defence of the common-wealth from foreign injury; and all this only for the public good.

In this single sentence, we find so many of the fundamental principles of conservatism in the US. Government should be limited in scope, and minimal in function. Its primary reason for being is judicial. It is there to resolve property disputes in a fair and unbiased manner in accordance with the laws that have been legislated through common agreements. Its second function is to provide a common defense, which is done through "force of the community." While this idea is certainly an anachronism in the face of postmodern warfare, it lingers, and is part of the reason that many conservative Americans fear having their guns confiscated or regulated.[1] This idea is further reinforced when we realize that Locke, again in contrast to Hobbes, believed that the social contract is one among equals who will remain equal after political power is established. This not a contract between ruler and ruled. Such a situation is tyranny, and is to be rebelled against. A state that usurps power as a mechanism for rulership cannot stand, and it would be within the rights of the citizenry to overthrow this power. This defense of liberty and citizen rights within the state, as well as the defense against usurpers from without, cannot occur without an armed citizenry. Such was US colonial revolutionary theory. Combine this idea with the belief that firearms were a fundamental tool of production and protection on the frontier, and it is very clear how conservative subcultures developed that are very attached to their weapons. Having these objects of production are worth whatever chaos or sacrifice must be endured. (CAE does also recognize a third component that is never spoken by the right, and that is the pleasure and intoxication that some people find in death and/or destruction. We need only look at any of the machine gun rallies documented on YouTube to witness a contemporary form of Dionysian revelry.)

Locke's Environmental Legacy

Again, we cannot blame Locke himself for the environmental disaster that his contribution to social thought bequeathed, because he could not know what was coming technologically or legally. However, the way in which conservatives (especially libertarians) still embrace Locke's ideas nearly verbatim (with the exception of the limits he set) is ruinous for the environment nationally and globally. CAE will briefly go down the list. First, people may do as they please on their own property as long as they are developing it (this is a particularly glaring problem in the US, seen in the 2017 devastation of Houston by flooding during Hurricane Harvey, due primarily to unrestricted, unregulated development). Second, the regulation of property erodes liberty on a secular level, and is an evil on the divine level as it retards providence. Third, all undeveloped land is wasteland. God has provided us with these resources and expects us to use them. Fourth, all land that is not in use is open for claim and development, and that includes all public land. (For many decades this was true in the US, as many public lands were exploited by ranchers and the extraction industries. In the early twentieth century, most of these practices were stopped or regulated, but the desire to reappropriate the lands never went away. It has long created furious anger among those who would benefit from a reopening of the land to agriculture and industry.[2]) Fifth, God has provided adequate resources that will last as long as humans inhabit the planet. Sixth, land development is providential in nature. Seventh, in this new chain of being (absent king and aristocracy), all nonhuman forms of life are "inferior creatures." This idea is not necessarily a problem when all creatures are viewed empathetically as life, but in the context of the Lockean world, where inferior status reduces a nonhuman creature to nothing more than a resource, it can only be problematic.

In addition to these problems, there are two large general ones that have emerged and are a direct danger to the environment: individualism and divinity. Individualism is the greater danger of the two when viewed in the Lockean context. Everything other than the ego represents only utility. Humans are collectively unimportant, and the fact that an individual may be a human is of no particular value or meaning. Individualism has contempt or at the very least indifference to humans, humanism, and, dare we say, anthropocentrism. Understood in this way, individualism represents an unempathetic and cruel aggregate of sociopaths (much like conservatives consider those who do not share their view of the world). This is where divinity comes in. The world is not a cruel place. It is a place of abundance where each individual has the means to care for him or herself. Human failure is due to sin conjoined with defective character traits. Humans do not need to take care of one another (no one is their brother's keeper), because God has freed us from that need. God is our shepherd, and we as individuals need only take care of ourselves both materially and spiritually.

Europeans are often mystified by why US conservatives are so tied to religion (and Protestantism in particular) in terms of governance and policy making, especially given America's constitutionally explicit separation of church and state. The reason is that they cannot maintain the plutocratic economic structure they desire without it. Without it, political positions such as anti-regulation (of anything, including the environment) and anti-welfare could not be justified. There is no other ideological justification that has any authority in US culture that would find such positions acceptable. American conservatives are by necessity ideologically stuck in the eighteenth century with no way out, much to the delight of the church industry. This is also, in part, why originalist interpretations of the constitution are so important to them.

In the end, this combination of individualism and divinity makes conservatives very suspicious of any system. Not just governmental or economic systems, but ecological systems as well. Rachel Carson convinced them of nothing. On the other hand, any anthropocentric individual (including some conservatives of a nonreligious disposition) will readily admit that if we want to keep humankind at the top of the food chain, preservation of the environment that allows for this status is a necessity.

The Anthropocentric Environmentalists

By the mid-nineteenth century in the US, the infant stages of a counterbalance to maximum growth at maximum speed in regard to use of natural resources began to surface. Proto-conservationist and philologist George Perkins Marsh (1801–82), who wrote the book *Man and Nature* (1864) and later redeveloped it as *Physical Geography as Modified by Human Action* (1874), was influential in changing the terms of environmental debate. His sophistication lay in the fact that he believed the romantic argument was untenable in such a pragmatic culture. Awe and ecstasy extracted from natural beauty were not of interest to most (and especially not to the powerful economic elites), nor was poetry a rhetoric of persuasion under these cultural conditions. Instead, he turned to economic reason topped with a rather stiff moralism. Those of the Lockean tradition understood this manner of speech. He was also willing to put humans at the center of the debate, but refused to do it at the level of individualism. For Marsh, consideration of what is good for all, as well as what is good for future generations, was fundamental to policy regarding natural resources.

In *Man and Nature*, Marsh argues with a clear subtext of (human) collective good, and goes so far as to say that without humans, the earth would settle into a state of senescence (clearly incorrect, but that is not of concern here). In certain conservationist moments, he does mention that conserving land would be good in that it would preserve indigenous plants and animals, but his primary concern is what makes a healthy environment for humans now and *in the future*. Marsh begins by redefining waste. He does not see it as the underdeveloped. Instead, he makes a proto-Taylorist argument: Yes, we want maximum production and profit, but we misunderstand what maximum efficiency is. What is being called maximum efficiency is actually wanton destruction. It is a product of immaturity, somewhat like children eating sweets until they are sick. A balance needs to be struck between business and stewardship. To clear-cut forests and leave a desert behind is not an efficient use of resources. Marsh was also somewhat of a historian, and believed that ancient civilizations had made this very mistake. This was an early argument for the importance of keeping reserves of resources and the development of means that allow continuous use of resources without ending in their exhaustion (in other words, sustainability). Until such measures are taken, Marsh thought we would only see the worst of humans through their destructive impulses. He very much tried to establish a moral high ground, and supported it through an economic argument about how people can get the most from forests, soil, and water.

The answer as to how to implement these ideas was through administration. Marsh himself was not a reformer, but his argument did win a considerable number of converts, and most significant were those in the Interior Department. Within this department, a political predisposition arose that aimed at reigning in frontier excess and environmental destruction, and began to think of environmental administration and use of public lands in a more future-oriented way.

Over the long term, Marsh's legacy is that he began a viable alternative argument to aesthetics and developed the foundation of the economic argument for conservation in the US. Like the aesthetic argument that is the heart of this book, the economic argument is just as profoundly anthropocentric. It boils down to the proposition that biodiversity and resource management are good because they increase the potential for products and resources to be harvested from the environment that will benefit humans in general (as opposed to just profits for a few). However, the argument that making more species available for exploitation over a longer period of time will fertilize the ground for a more robust economy for all was no doubt distasteful to the more environmentally liberal-minded of the time period. They were not of the belief that the economy trumps humanity's higher callings and qualities regarding relationships among living things. Unfortunately, within capitalist society, where economy comes first, this rhetorical form of economic argument will not go away (and has not). Environmentalists have to fall back on it to one degree or another, and it is probably the only rational means to persuade neoliberals to change course in regard to the environment.

In spite of the growth of bureaucratic mediation between labor and nature, romanticism did not disappear. To the contrary, it began to organize, and turn its philosophical position into policy. Enter environmentalists' favorite writer, naturalist, and activist, John Muir (1838–1914). He and friend and colleague Joseph LeConte would bring the poetry back, and defend the perspectives of Walt Whitman, Ralph Waldo Emerson, and Jean-Jacques Rousseau, which represented a continuance of the idea that nature has an underlying moral harmony, and that one learns morality through communion with nature. But what Muir and LeConte were most concerned with was aesthetics (the scenic). Indeed, this is what makes them so deeply anthropocentric. For Muir, human experience is fundamentally different from that of the creatures who inhabit the wilderness. In nature, and especially in spectacular places of perfect form such as the High Sierras, people can have a totalizing aesthetic experience—a feeling of holiness, of ecstasy, of awe, or of the sublime. Areas most inclined to produce this effect were the places Muir and the Sierra Club sought to protect most. An aesthetic hierarchy was produced as a means to choose which lands should become parks and/or, later, monuments. Human experience was the key to the hierarchy. The mundane landscape was of little interest; rather, it was the places of grandeur that were to be protected, the places where people could look at nature in its most powerful form in terms of its effect on humans. In his case for protected public lands, Muir catapulted the idea of the importance of looking at nature (preferably directly), a tradition that continues to this day, and around which a huge industry has been built in direct tourism and in media that use nature as their subject. His advocacy for sending people into nature for intense and overwhelming experiences far beyond those of everyday life was a smart tactic in building a public will to protect the parks from the business interests that wanted that land back in production. The deep emotional bonds people built with the parks made them politically dangerous to disturb.[3]

Like all good romantics, Muir was a man who could live with personal contradiction in a manner that made sense. Muir was very fond of the idea of powerful individualized experiences of nature, and believed that it was a good idea to seek solitude in one's communion with nature, but part of the reason for this was so that individuals could realize how small they are—what a tiny, finite spark a person is in this great cosmos. Such profoundly individualized experience ironically deheroicizes the individual, but that is not the big contradiction. More significantly, Muir believed that communing with nature is important in order to build "fraternity." Experiencing nature together with others is very significant on a number of levels. First, it builds or strengthens the bonds of love and friendship among those who experience nature collectively. Second, it helps people to understand the public good and the value of collective investment. That one and all own these cathedrals of nature and that we may experience them at will demonstrates that what is good for one can be good for all in a manner that creates a transcendental shared happiness, rather than a separate, parallel happiness that the individualist would associate with public affairs.

Muir does not stop there. We remember that in the Lockean universe, people are measured by their relationship to labor (the production of value). Muir would have none of this. While certainly not a believer in zero work, he questioned whether labor is the core of a person's character and whether the surplus value produced through work is really any kind of luxury. For Muir, the highest state of being is total aesthetic experience, and that can only be acquired in nature—not at work. Muir viewed the Puritan work ethic as the recipe for a mundane life. He called for vacations so people could pursue higher callings in nature. He saw the nose-to-the-grindstone attitude as counterproductive to a person's humanity. It was time to stop working so much and have some fun; give up the mundane for the exotic; forget the conflict of toil and embrace nature's harmony; and do so with family, friends, and other enthusiasts to create emotional bonds, rather than the utilitarian relationships and associations that come with work.

Muir attacked the Lockean world on all fronts. Nature is good and complete in and of itself, and should be experienced in this form. Individualism and parallel happiness is not as important on human grounds as the aggregate happiness produced in community. Labor is not the measure of a person, nor does it create riches. At best, labor helps us produce the value necessary to allow us to take time in nature to enjoy aesthetic pleasure and higher human pursuits. Muir is preaching not just a worldview, but a whole new lifestyle. The lesson that Muir seems to be teaching is that we conserve, not because it is good in and of itself, or because, as with Marsh, we need to Taylorize our resources; we conserve because it is good for *people*, as that is how we discover the higher purpose of our own humanity.

Now to an even bigger contradiction, and a very strange bedfellow for environmentalists: Theodore Roosevelt (1858–1919). A staunch individualist who thrived on personal achievement and glory, he advocated for what he called a "strenuous life" in which individuals would repeatedly seek the greatest challenges to individual survival and steadfastness. Among his favorites were war and, perhaps second to that, big game hunting. These two forms of adventurist pleasure also brought twinges of anxiety for Roosevelt. Since there was not enough war for youth to test themselves, something had to take its place. For some it would be American football—a game that, at the time, had few rules and was ultraviolent, typically with serious injurious consequences that were sometimes deadly (the worst year was 1905, when nineteen deaths occurred). Roosevelt himself had to intervene to lower the intensity of the violence. For Roosevelt, if war was not an option, the second-best situation for the development of character was answering the call of the wild with weapon in hand for confrontation with the savage beasts of the wilderness. Thus, while he was worried about the proper socialization of the American male, he became equally concerned about the need for wilderness as a site where proper stress could be applied to youth. Not to mention that he and his comrades at the Boone and Crockett Club (a group he founded in 1887 consisting of one hundred economically elite conservationists with an interest in hunting) wanted wild areas to hunt in as well. This predisposed Roosevelt to support public lands and parks. And he figured out that, as president, he could order the conservation of huge allotments of land as national monuments. He went on to do what Marsh had suggested many decades earlier: manage the parks through administrative institutions with tremendous power to act independently in setting policy for the lands (politics without politics). So while some of the lands would continue to be worked, they would be worked as renewable resources, and environmental devastation would stop. Some lands would remain fully protected, and more wilderness areas would be designated as parks and monuments for their protection.

Herein lies the contradiction. While Roosevelt was an advocate of individualism, he also understood the need for collective investment, and saw public lands as an excellent form of public investment. Facing the savage beasts aside, Roosevelt viewed communing with nature as a social good where people build bonds with one another and experience our connectedness, "fellow feeling," and brotherhood. He saw nature as a place where social differences in conflict can play out in a positive way through shared recognition of common ground. At the time, he was particularly concerned with class conflict. At this point in American history, the distribution of wealth in favor of the rich was greater than it had ever been, or would be, until present-day America. Roosevelt believed that the parks and monuments, these common shared lands, could bring the classes back together and lessen the animosity between them. To his mind, while investment in public lands as a vehicle for individual expression or as a conservationist gesture was important, investment in scenic wilderness for its social value was equally so.

Roosevelt's bureaucratic paradigm of conservation and land management remains in place to this day (although it is under heavy attack), and the Rooseveltian ideology of conserving nature remained until the 1960s, when a new ecological perspective emerged.

In the early 1960s, people with environmental sympathies were captivated by the systems theory of ecology, best expressed to the public imagination through Rachel Carson's *Silent Spring* (1962). Carson (1907–64), a marine biologist and conservationist, may not have been the first to speak about nature as a flowing system of interconnections—that honor probably goes to Fairfield Osborn and his work *Our Plundered Planet* (1948)—but she was more convincing. She found the mechanism that spoke not only to liberal humanists predisposed to an attraction to nature, but to middle-class suburbanites who did not have the fate of humans on this planet on their lists of major concerns. Her explanation was concrete and expressed clear and immediate danger for all. Her description of the flow of DDT through the ecological system made a compelling case for the profundity of the concept of nature as an interconnected and interdependent system. The movement from air, water, and soil, to the plants and animals of land and sea, to the human food supply, and finally to our own bodies made an impression. If the planet was being poisoned, so too were people. Humans are a part of the system even if we (believe that we) have dominion over it. Should the ecosystem collapse, humans will die with it.[4] If a reader only gets through the first chapter, "A Fable for Tomorrow," the core of the persuasion is what a planet of toxins will mean for humans, rather than the argument that not polluting the planet is good in and of itself. While *Silent Spring* can be read as a document that challenges human centrality in the world, and attempts to nudge it toward a more marginal position, a reader can get through this book with anthropocentrism intact, and all is well and good, as the constituencies who were needed to see that DDT was banned were not alienated. The slogans inspired by her work to eliminate DDT were not "save the song birds" or "keep our aquifers pure;" rather, they insisted that this must be done to save the children and provide them with a habitable world in which to mature.

CAE must also reiterate that there was a sizable population that Rachel Carson did not convince of anything. Those who were still committed to the colonial ideology of the eighteenth century remained as suspicious as ever of systems, and believed that maintaining them overdetermines the actions and freedoms of individuals. Why attack herbicides and pesticides when weeds and insects must be controlled, and when controlling them makes yields and profits higher, and expands job opportunities? For this demographic, regulation of industry is never a good idea, especially when driven by ecological alarmism.

Anthropocentrism is not necessarily the enemy, and has in fact enabled healthier forms of necropolitics as well as environmental consciousness itself. Historically, it has been a key element in the persuasive rhetoric of those who truly care for the environment, and it has helped their arguments resonate with potential allies. Anthropocentrism and anthropomorphism are the foundation of empathy, connectedness, and investment in the natural world. To be sure, they are not useful concepts for scientific study, but they are of great use in poetics and aesthetics. Much of ecological struggle is being fought in this nonrational territory—which leads us next to ask: Is there a human tendency toward the nonrational?

Notes

1. In the South, guns were also considered a necessity by slave owners and supporters of slavery in order to maintain mastery over slaves and to put down slave revolts. After slavery, the residue of this practice transformed into a means for race control, and finally into stockpiling weapons for an upcoming (mythic) race war. These emerging post-slavery practices transcended regional concentration, and became lightly distributed through-

out the nation. Presently, most conservatives appear to be indifferent to the former, and only a small minority is participating in the latter.

2. By way of example, consider the two Bundy uprisings. The first was in 2014. Nevada rancher Cliven Bundy had refused to renew his license to graze cattle on federal public land in 1993. In 2014, the Bureau of Land Management (BLM) obtained a court order directing Bundy to pay one million USD in back fees. Bundy refused to pay, saying he had an inherited right to use the land. The BLM closed the allotment that Bundy was using and began to round up his trespassing cattle. At one of the cattle gatherings, Bundy and his supporters, who included sovereign citizens and various militia groups (many of whom were armed), showed up to reclaim the cattle. In order to de-escalate the situation, the BLM agreed to release the cattle. Bundy still grazes his cattle on public land, has not paid back fees, and has not renewed his license. It is amazing to think how perfectly Locke's ideas on property rights in regard to public land, the right to rebel against oppressive rule, and the inherent mistrust of regulation and administration have been perfectly preserved, even though this form of land use interpretation for public lands has been relatively dormant for the past century.

The second uprising occurred in 2016 when Ammon Bundy (Cliven's son) and supporters from local militias and sovereign citizens' groups occupied the Malheur National Wildlife Refuge. Their primary demand was for the refuge to be returned to the state of Oregon, and completely taken out of federal hands. This was in part a continuation of the Sagebrush Rebellion, in which Utah senator Orrin Hatch attempted to pass legislation that would limit designation of land and wildlife protectorates (although since the election of the Trump administration a form of this bill is back). The bill failed, but the Reagan Administration slowed the designation process considerably. The ultimate goal of the uprising and the legislative action was to return as many public lands as possible to state control, where ranching, extraction, and other industries would find a more sympathetic and cooperative political system.

3. Americans' love of their national parks and monuments is about to face its biggest test ever from the greed of business interests and their supporters. At the writing of this chapter, the test case is Bears Ears National Monument in Utah, where the right hopes in an unprecedented move to rescind its monumental status in order to allow further development, primarily by the extraction industries. The monument is very new (created by

Obama in his final days as president) and notably not a park. It is located in a deeply conservative state as well as a site that has been heavily contested for decades. In a mixed blessing, the monument status has not been rescinded, but the acreage is going to be reduced. We do not yet know by how much, or which industries will able to access the newly unprotected land. No doubt considerable blowback is on the way, and this order will certainly end up in court. Other parks and monuments appear to be safe at the moment, but should this test case work out, more attacks on parks and monuments should be expected.

4. Carson managed to make a major contribution to transforming environmental stewardship into a nonpartisan issue for the majority of people in the US; this would continue until Reagan, and then deteriorate rapidly into the merciless partisan warfare over the environment that exists in the US today.

There Will Come Soft Rains

(War Time)

There will come soft rains and the smell of the ground,
And swallows circling with their shimmering sound;

And frogs in the pools singing at night,
And wild plum trees in tremulous white,

Robins will wear their feathery fire
Whistling their whims on a low fence-wire;

And not one will know of the war, not one
Will care at last when it is done.

Not one would mind, neither bird nor tree
If mankind perished utterly;

And Spring herself, when she woke at dawn,
Would scarcely know that we were gone.

–Sara Teasdale, 1918

3

Schopenhauer Today

After a multi-decade commune with Marx and Hegel, Max Horkheimer returned to the study of one of his early philosophical interests, Arthur Schopenhauer (1788–1860). By the 1950s, Horkheimer was distraught over the immense amount of suffering that had occurred in the first half of the twentieth century. Hegelian optimism made no sense given the historical record. The idea of progress was broken and abandoned. The sociologist Pitirim Sorokin summed up the mood well in *The Crisis of Our Age* (1941) when he stated, "In few periods of human history have so many millions of persons been so unhappy, so insecure, so hungry and destitute, as at the present time, all the way from China to Western Europe." A strong sense of pessimism filled the air. Five decades of war, depression, genocide, massive class inequality, failed revolutions, emergent authoritarian states (on both sides of the political spectrum), and potential global destruction from atomic war left Horkheimer concerned for the future, but more significantly, wanting to understand suffering—a topic of some significance in the philosophy of Schopenhauer.

We might all do well to follow his example, as the suffering is still as great, and the many crises in the economy, politics, and the military continue unabated in the twenty-first century, with environmental and ecological crises including climate change, mass extinction, the poisoning of the planet, and environmental injustice added to the list. Pessimism still hangs in the air, but CAE has some additional reasons to check in with Schopenhauer. We want to return to Schopenhauer to examine the centrality of the nonrational in his philosophy, as well as his conceptions of nature in all their glorious contradiction. We also believe his thought might shed light on the role of art within the frame of environmentalism. Through the work of the original curmudgeon, and the first to argue the opposite case whenever people or civilization became too self-congratulatory, we may be able to confront some less-than-pleasant thoughts that emerge when we think about environmental struggle and justice.

Kicking Against the Pricks I: People, Politics, and Antihumanism

CAE admires Schopenhauer's contrarian tendency (in fact, it is a trait we hope to emulate in this very book). Wherever he saw philosophical consensus, he was sure some major delusion was behind it. Consequently, Schopenhauer introduced a number of heretical ideas that opposed centuries of consensus in Western philosophy. One of these notions was that humans are primarily nonrational in temperament and behavior. The common notion at the time was the Socratic one, in which reason drives the human chariot, keeping the horse of will and the horse of desire in check and on the path to wisdom. Schopenhauer believed this to be nonsense. Indeed, humans have a capacity for reason, but they do not use it to pursue lofty goals (if such things occur, it is incidental to the actual motivation). Humans are driven by innate needs and desires. Drives are nonrational and unaware of either outer or inner experience. They are simply blind forces that push the individual ever onward. Reason serves to fulfill these needs and desires, although it must navigate social convention and law, which act as constraining exterior forces. In spite of these constraints, people, places, and things are resources to be used to the best extent possible to suit individual needs. They are, in a word, utility.

Schopenhauer was also heretical in that he viewed humans as overdetermined and far from being masters of our own fate. Not only are we at the mercy of drives we do not understand, and that we often do not even admit exist, but our behavioral tendencies are determined by our character. Schopenhauer believed that each person has a set of essential traits that set strict parameters on their thinking and behavior. These traits do not change over time, and continuously act as a limiting force in terms of behavioral options.

The idea of humans as nonrational and overdetermined does have a legacy. In the early twentieth century, this point of view was the foundation of psychoanalysis, whereas today it is theorized in behavioral economics. Behavioral economics is a discipline combining concepts and data from economics, social psychology, sociology, evolutionary biology, and neuroscience. It has two fundamental insights: that humans are not very rational when making economic decisions, and that humans are not entirely in control during decision-making processes. The discipline is particularly sensitive to what it means to make this kind of suggestion. Practitioners know that their work challenges the institutionalized, dominant microeconomic view of humans as rational actors exercising free choice (although the literature on behavioral economics and macroeconomics is growing every day). While they are careful not to say that humans are irrational, they argue that we are using or being driven by other factors than reason and data (and not always with terrible results). Interior determining factors in decision-making include instincts, emotions, aversions, and personality traits (along the lines of what Schopenhauer meant by "character"). Accessibility also plays a key role, which may mean accessibility of data, but may also mean emotion or hearsay that can crowd out any rational resource. Humans tend to be lazy decision-makers in so far as the force that is most convenient to them will most heavily influence a decision. Exterior factors include social consensus (which quite likely is not tethered to the real), opinions of friends and family, social convention and habit, and personal history. This latter category is very broad and would include everything from all types of trauma to prior commitments (people tend to stay with what they know). Behavioral economists are delivering a gold mine of resources to help marketers better exploit the deficiency of reason that humans suffer, and for us to better understand how we actually act in the world.

Returning to Schopenhauer, this conception of humans led to a fundamental form of antihumanism. His was the genuine article. CAE is not talking about some postmoderns labeling themselves as antihumanist because they object to the Enlightenment model of the universal subject that marginalizes or completely fails to recognize the vast majority of humans in the world. Schopenhauer's was a deep indictment of the uselessness of humans and the meaninglessness of humanity. Humans are nonrational, amoral, conflicted, miserable, delusional egoists who are moving ever closer to extinction. The German philosopher Ludwig Feuerbach once said about Christians that for them, "the highest degree of illusion comes to be the highest degree of sacredness." This quote was echoed by Guy Debord, referring to deceived, misguided individuals pacified by spectacle. If this quote had a meaning to Schopenhauer, it would be about human illusions in regard to our own nature and state of being. We think of ourselves as reasoned beings asserting free will in a field of ubiquitous choice in an atmosphere of social liberty, when for Schopenhauer, nothing could be further from the truth.

Schopenhauer's antihumanism brought him directly to conflict theory when it came to political thought. His interest was in English social philosopher Thomas Hobbes (1588–1679). Hobbes postulated a presocial state marked by conflict, in that all are sovereign, so all may do whatever they please when pursuing their interests. For Hobbes, it was a situation of tremendous disorder, if not total anarchy. Hobbes referred to this state as *the war of all against all*. Schopenhauer liked this concept very much, but disagreed vehemently with Hobbes on one matter. Schopenhauer believed that Hobbes was not describing the presocial state, but the current and permanent state of human interaction. The social fabric was little more than numerous egoists violently bumping up against one another, constantly frustrated, and in various degrees of psychological and/or physical pain.

CAE can only guess what Schopenhauer thought of the idea of democracy, but based on his few writings regarding social thought, we speculate that he thought it so silly that commenting on it would be beneath him. Since democracy requires its participants to be rational enough to understand what is best for them, through reasoned discourse to persuade others it is in their interest too, and furthermore, to show how it could be implemented in policy, it seems very likely that Schopenhauer would only find this laughable. Creating a governmental system with a backbone of rationality, reason, and informed participation was something far beyond human potential. Governments existed only to facilitate egoism in all its worst forms. Not that Schopenhauer was against a state attempting to maintain order amidst the sociopaths who populate it. He was a man of means and lived in constant anxiety that he might lose it all in a revolutionary (anarchic) moment.

Given the wretched state of humans, humanity, and society, what should a person do? Schopenhauer's answer is resignation. Retreat to where one must in order to find some semblance of tranquility through the examination of inner experience, and wait out the punishment imposed upon the living for no reason or purpose. Disengage from the battle that is ordinary life, for the harder we fight the forces that drive us, the more intensely those same forces assert themselves.

For CAE, this is the endpoint of antihumanism—retreat and disengagement. Why would we want to save an environment any more than we would want to preserve the instruments of our own torture? In Chapter One, we saw the active endpoint of green antihumanism: intensify the chaos, and stimulate and accelerate the forces that push humans in directionless and meaningless capacity until humans meet their own destruction. Schopenhauer presents the passive alternative. Just wait it out. Humans are busy plotting and enacting the destruction of the human race and the planet at this very moment. Why join the frenzy? And in the end, what will it matter? Humans are just another product of the individuation of the Will. Their destruction is of no consequence, either for good or bad.

CAE does not see either of these alternatives as viable ways to live. Even if in the deepest, darkest recesses of our minds we might believe Schopenhauer to be correct, we would rather place our bets on the hope that he is not. Even if we use Schopenhauer's own method for understanding the order of things, and use our inner experiences to gain a better understanding of what is hidden in outer experience, we do not find the level of torture that Schopenhauer says we suffer (although we are certain that there are many in this world who do), nor do we find outer experience to be of a hellish fury that would rival Dante's *Inferno* (although we know of places on the planet that are). We are of the suspicion that Schopenhauer is indulging in some exaggeration. Moreover, while CAE does recognize biological, psychological, social, and economic constraints on individual liberty, choice appears to be ubiquitous.

Kicking Against the Pricks II: The Two Natures

Schopenhauer's second great contrarian moment lies in his thoughts about nature, but to understand this we will need to briefly review his metaphysics. Schopenhauer is intensely interested in the dual aspect of existence. The center of understanding this situation is the body. We can view our bodies objectively as representation or as outer experience, but are also able to view them subjectively as inner experience. We can access the drives, feelings, perceptions, thoughts, and physical processes inside our own bodies. The body is unique in that it is the only object that can be known in this dualistic manner. All other objects can only be apprehended as representation. For Schopenhauer, we are fortunate in that we only need this one example to understand reality and appearance. Schopenhauer was always very concerned that Kant had given too much to skepticism in concluding that the thing-in-itself is unknowable. Schopenhauer thought that our inner experience gives us access to the real. The real is a singular thing—the force that courses through us and by necessity all else that appears. We come to understand this force through our instinctual drives. This force that manifests as the plurality of objects represented in the world he calls Will. (He does this because it makes the concept easier to understand; since we can comprehend it through inner experience, it is less vague than saying there is a singular energy, force, power, or other term commonly used to describe the thing-in-itself.) In *The World as Will and Representation* (1818, 1844), he describes it this way:

> Just as a magic lantern shows many different pictures, but it is only one and the same flame that makes them all visible, so in all the many different phenomena which together fill the world or supplant one another as successive events, it is only the *one will* that appears, and everything is its visibility, its objectivity; it remains unmoved in the midst of this change.

Within this system is the contrarian element. Schopenhauer believes that the Will is not divine, but nonrational, blind, aimless, and void of intent or compassion. How it manifests as representation is accidental and meaningless. Not that he has a disregard for science (although he emphatically describes what he believes to be its limitations)—he understands the planet as having order in terms of physical laws or chemical reactions, but it is not an intentional order. There is no God, Great Architect, or Divine Designer. Teleology is a myth; complexity or expanse does not impress him. Nothing has purpose. Natural process is meaningless. It is blind, groping, and its products may flourish or die off. It really does not matter. Whether we are in the social world or the natural world, everything is oscillating between order and chaos, emergence and decay, life and death. In its blindness, the Will has no concern for contradiction, disharmony, or brutality, so once the Will is individuated and objectified it turns upon itself, fighting itself, consuming itself, destroying what it creates. Nature, like society, is no moral teacher. It has no relationship to good and evil. All it does is continue, to no end in particular. How and in what form is irrelevant to the Will.

This is a very modern view of nature. Indeed, Schopenhauer may be the first modern to propose this now all-too-familiar view. He is a proto-evolutionist, and is laying the foundation for the scientific view of nature. While he could not have been familiar with key scientific principles such as selection, mutation, or adaptation (because these came after his time), he did understand nature as a completely secular process that has no meaning beyond its immediacy. This declaration from the early nineteenth century posed a profound choice that haunts us even now: we must side with the mystics and accept divine intervention and intention, or we have to answer some very difficult questions regarding meaning, ethics, and actions. In terms of environmentalism, we have to answer Schopenhauer's suggestion of resignation in the face of a scientific view of the natural world. Why should we care? If life is meaningless, why should it be valued?

Today, we know that the fate of every species is to be eliminated. If it is fortunate, a species may last four or five million years. During that period, it is likely that either the environment will change beyond the species' ability to adapt, or a better-adapted competitor will emerge, or the species will evolve into a different one. Scientific estimates are that 99 percent of the species ever to exist on earth are now extinct, and that much of that extinction was due to bad luck rather than competition from a better-adapted species. Who knew a massive meteor would hit the earth? There have been numerous times on the earth when there were relatively few species. It may be comforting to think that evolution is working from the simple to the complex and from singularity to great diversity, as that is how we like to think about the division of labor and culture, but in reality, evolution represents an oscillation pending conditions on earth rather than unilinear or teleological development. Science gives us no reason to prefer biodiversity over a small aggregate of species, or to prefer a mammal over bacteria or vice versa (although from an anthropocentric position, these choices are easy to make). Moreover, earth itself has an expiration date when everything will die. Even if luck is with the planet and there are no cosmic accidents that may cause this end-time calamity prior to the sun becoming a red giant, we know an end is coming. Even more chilling is the reasonable probability that the cause will not be cosmic ill luck, but rather human error that leads to a fatal global accident.

The associated preference for "sustainability" is also delusional when placed within a system that is clearly finite. Nothing is sustainable in an ever-changing world. This term can have economic pragmatic meaning, or it can disguise the desire to keep humans at the top of food chain, but it is meaningless in terms of evolutionary process. It also reveals very arbitrary or expedient choices among temporal units. When we are thinking environmentally, are we thinking with urgency, or about a decade, a lifetime, a century, a millennium, evolutionary time, or geologic time? The answers to what is sustainable and how to practice it vary with this choice. Within this system, can we even make the claim of an "ecological crisis"? Is that not anthropomorphizing in a way science would find unacceptable?

In the end, arguments concerning environmental crisis can only come from the arts and humanities. Science can tell us with modest assurance the probability of future occurrences within the environment, why they are likely to happen, and perhaps what can be done to change them, but only the arts and humanities can call them a crisis, or label the occurrences as positive or negative. Once scientists do that, they are out of the realm of science. Protecting the environment is only arguable from a humanistic, if not anthropocentric, position. Otherwise we are obligated to be mere observers (perhaps with compassion, but without intervention), watching a lion kill the cubs of its rival.

CAE has promised two natures and a compelling contradiction, and we intend to deliver just that. As noted above, we tend to think of Schopenhauer's version of nature as one of tremendous cruelty, where the many manifestations of the Will collide in brutal scenes of lust and violence without a trace of mercy or pity—a position that would inspire the naturalist and decadent literary movements in France later in the century. However, Schopenhauer had a soft side—times when his pessimism abated, and his flair for seeing suffering in all manner of life was subdued. One place was in art, and the other was in nature. In one moment nature is a scene of merciless brutality, and then, in an almost Jekyll-and-Hyde manner, nature becomes a wonderful place. Consider the following:

> Yet how aesthetic nature is! Every little spot entirely uncultivated and wild, in other words, left free to nature herself, however small it may be, if only man's paws leave it alone, is at once decorated by her in the most tasteful manner, is draped with plants, flowers, and shrubs, whose easy unforced manner, natural grace, and delightful grouping testify that they have not grown up under the rod of correction of the great egoist, but that nature has here been freely active. Every neglected little place at once becomes beautiful.

Passages such as this one are not few and far between; they are sprinkled throughout both volumes of *The World as Will and Representation* and function as modest relief from the grim elements that populate most of Schopenhauer's works. Apparently, natural beauty has powers to soothe the savage pessimist, leading to the odd appearance of what reads like romanticism. Most assuredly, these lines could have been written by John

Muir himself. Schopenhauer shares with Muir the belief that humans have a different relationship to nature than do other beings who inhabit it. We can get something from nature that other species cannot—the highest form of human experience—totalizing aesthetic experience that functions as pure knowing. Nature serves us in a threefold way. First, and most importantly, nature in her profundity, great and small, has the potential to disintegrate ego. We can lose ourselves in nature and achieve a temporary disintegration of instinct and desire through our vision of nature's grandeur or beauty. Second, this process is pleasurable even when it is also terrifying. Finally, once returning from this higher state, we come back with a greater understanding of the real. We cannot know if Schopenhauer would have been a preservationist, but in passages such as the one quoted above, he certainly sounds more like a romantic than a pessimist, and indicates that nature should be left to itself. CAE's perception of Schopenhauer's more romantic side comes from our reading of his argument that nature can be most beneficial for humans by providing the great luxury of aesthetic experience (something woefully and painfully missing from everyday life in Schopenhauer's view). As a way to close, here is Schopenhauer's description:

> If we lose ourselves in contemplation of the infinite greatness of the universe in space and time, meditate on the past millennia and on those to come; or if the heavens at night actually bring innumerable worlds before our eyes, and so impress on our consciousness the immensity of the universe, we feel ourselves reduced to nothing; we feel ourselves as individuals, as living bodies, as transient phenomena of will, like drops in the ocean, dwindling and dissolving into nothing.

Kicking Against the Pricks III: Art and Aesthetics

Arthur Schopenhauer was a philosophical superstar of the modern art world. He may have had little influence in any discipline in the course of his lifetime (although there was a small popular discovery of his work in philosophy at the end of his life), but he more than made up for it after his death, as he was incredibly popular in the 1870s and '80s and, while less so after that, was still influential until the 1920s. CAE would say that the peak of his appreciation was akin to the popularity of Jean Baudrillard in the 1980s, or Gilles Deleuze in the 1990s. To be a part of the cultural discourse of those decades, one had to be at least conversant in these theoretical brands. In the late nineteenth century, whether one was a composer, a visual artist, or a writer, Schopenhauer needed to be a part of one's world. Schopenhauer was a gold mine of ideas, so people took what interested them, and often for opposing purposes. The literary naturalists were fascinated by his concept of Will, through which Schopenhauer read nature in a more pessimistic way. At one point Émile Zola became so enamored with Schopenhauer that he wrote a novel, *The Joy of Living* (1883–84), illustrating the philosophy. The decadents were interested in the quest for aesthetic experience, and how this quest could become a lifestyle in all its final tragedy—the finest example being J. K. Huysmans's *Against the Grain* (1884)—another true homage to Schopenhauer. As decadence evolved into symbolism, aesthetics became the total interest, along with a few of the more occult elements. And there were those, such as Richard and Cosima Wagner, who simply accepted Schopenhauer as the modern gospel. Cosima Wagner often featured readings of Schopenhauer in her salon, which was nothing less than a showcase of the great scholars and artists of the time.

Perhaps Schopenhauer simply fit the mood of the time, as one and all seemed interested in the tragedy of life. Or perhaps it was because, in him, artists had something they had never had before: a champion. The old contrarian had done it again. He was willing to argue that if humans wished to experience and understand the real both as Will and as (Platonic) Idea, the means to achieve this objective was art (although as we have noted, nature could be a means as well), and that the noblest of dispositions was the artistic one. Philosophy, and metaphysics in particular, was next in the hierarchy, as it could explain *why* this is the case. Bringing up the rear were the sciences, which offer only practical information. This was quite an unexpected argument coming immediately after the Enlightenment. Consider the following:

> Whilst science, following the restless and unstable stream of the fourfold forms of reasons or grounds and consequents, is with every end it attains again and again directed farther, and can never find an ultimate goal or complete satisfaction, any more than by running we can reach the point where the clouds touch the horizon; art on the contrary, is everywhere at its goal.

Much like nature, Schopenhauer begins with the idea that art is judged by its effect on the subject. We know we are witnessing art when we can access the essence of things. It transforms us from individuated consciousness to universal consciousness, which in turn makes possible the apprehension of the Ideas. Schopenhauer states it as follows:

> But now, what kind of knowledge is it that considers what continues to exist outside and independently of all relations, but which alone is really essential to the world, the true content of its phenomena, that which

> is subject to no change, and is therefore known with equal truth for all time, in a word, the *Ideas* that are the immediate and adequate objectivity of the thing-in-itself, of the will? It is *art*, the work of genius. It repeats the eternal Ideas apprehended through pure contemplation, the essential and abiding element in all the phenomena of the world.

Schopenhauer goes all in with Platonism with his ideas on eternal forms (which is ironic, since he vehemently disagrees with Plato's thoughts on art, calling them among the worst philosophical mistakes in history). His notion of the Ideas maps perfectly onto Plato's thoughts about them. If we take the example of the earth revolving around the sun, Schopenhauer would claim that this is temporary, because eventually the relationship between the planet and star will end. However, the *form* of two spheres in an elliptic relation will last forever. The idea is outside of time and space. When an artist paints a flower, he (Schopenhauer recognized only male artists) will be able to see and capture the essence of the flower, its Idea. That is because his genius is able to capture "not what nature has actually formed, but what she endeavored to form." The artist-genius has a peculiar way of seeing beyond the contingent, and this is the talent that makes him a genius.

From both art and nature we are able to apprehend both beauty and the sublime. The effect of a beautiful object of contemplation upon the viewer is to free the intellect from its service to the Will, allowing the self to wither away and leaving the viewer as a pure subject of knowing. Equally important to Schopenhauer is the affect that emerges in this state—one of intense "exaltation." Happiness, pleasure, knowledge, tranquility—for Schopenhauer these are all things positive, all things that negate the Will in both objective and subjective forms. Schopenhauer describes the effect as follows:

> By calling an object *beautiful*, we thereby assert that it is an object of our aesthetic contemplation, and this implies two different things. On the one hand, the sight of the thing makes us *objective*, that is to say, that in contemplating it we are no longer conscious of ourselves as individuals, but as pure, will-less subjects of knowing. On the other hand, we recognize in the object not the individual thing, but an Idea . . .

Objects that assert themselves as art, but are not, are easily discovered because they do the opposite—they excite and agitate the Will. They draw us into the subjective (the feeling of the sensual), thus reinforcing our ordinary disgraceful state of egoism, and immerse us in the individuation of the world. A well-executed painting of a nude will be exalting, while a poor one will be exciting. Equally as guilty of base sensuality are objects of charm and of negative charm (the disgusting—not to be confused with ugliness, which has a place in art), which should be avoided at all costs.

The sublime contains the beautiful and it, too, consists of phenomena that quiet the Will, in spite of the fact that the sublime is complicated by threat (physical and/or psychological). One might think that threat would inspire fear that would agitate the Will, but Schopenhauer believes that in the face of the sublime in all its various intensities, the opposite occurs. In a state of aesthetic contemplation, tranquility prevails (so long as actual immediate harm does not occur). Again, Schopenhauer:

> The storm howls, the sea roars, the lightening flashes from black clouds, and thunder-claps drown the noise of storm and sea. Then in the unmoved beholder of this scene the twofold nature of his consciousness reaches the highest distinctness. Simultaneously, he feels himself as individual, as the feeble phenomenon of will, which the slightest touch of these forces can annihilate, helpless against powerful nature, dependent, abandoned to chance, a vanishing nothing in the face of stupendous forces; and he also feels himself as the eternal, serene subject of knowing . . . he himself is free from, and foreign to, all willing and all needs, in the quiet comprehension of the Ideas.

The majority of Schopenhauer's examples of the objects of sublime experience are from nature, and understandably so, given the need for such massive scale. He tries to argue, often unconvincingly, that architecture from antiquity such as the pyramids could have this effect, but examples from nature are his preference. The type of art that is best suited to the sublime is tragedy, which partially explains the late nineteenth-century European writers' fascination with tragic tales.

Although aesthetic experience sounds like redemption, Schopenhauer is not willing to go that far. These experiences are short-lived. Dandies may try to make permanent aesthetic experience a lifestyle, but they will fail, and the failure will be tragic. The Will snatches a beholder back into the individuated world as fast as a demon will return an escaped resident of hell back to the boiling pitch. For Schopenhauer, resignation and asceticism is the better long-term strategy, imperfect though it may be.

Our Turn to Kick

At the end of this reassessment of Schopenhauer, CAE, like Horkheimer, thinks that he still has something to teach us, and that he will remain in the philosophical canon even if humanists cannot bear the darker side of these teachings (while also fearing in the back of their minds that he might be right).[1] The most significant lesson is to not underestimate the power of nonrationality. The Al Gore tactic of making a reasoned argument supported by data is simply not enough to convince most people of anything. Nor will scientific consensus, because huge numbers of people have no idea what that means. The great weakness of the humanities and sciences in terms of the environmental crisis is that we expect rational actors in rational systems. Argument has to be augmented with strategies and tactics that mimic the insights into nonrationality of behavioral economics, marketing, and advertising, in order to relay "small-*t*" truth through means other than reason, and ultimately to bring the participant back to a position of reason. (This is what distinguishes interventionist practices from propaganda, which relays falsehoods through a battering of nonrational qualities). Searching for the "bliss point" is as significant as mining the data. In our more performative projects, CAE has consistently sought out methods that allow us to translate the scientific consensus into more consumable but less aggressively ideological packets for nonspecialized audiences.

Over the years, CAE has also had an interest in qualitative microsociology, and has put some effort into our admittedly amateur studies of people who do not believe in environmental crisis, or are indifferent to it. The population we are speaking of here excludes the cynics who deny climate change because they can profit directly from doing so. For various reasons, CAE gets to spend time outside our bubble with people who deny environmental crisis or do not care about the environment, and not because they do not have a deep appreciation for nature. Most do, and spend a great deal of their time in nature (a good deal more than most urban dwellers). These are folks who cannot be convinced of much by argument or study. They are convinced by their own experiences in the world, which are then reinforced by family, friends, and neighbors. When they go outside, they experience clean air and water, and an exuberant nature that is carrying on as it always has. They register no sign of crisis, so environmental issues are not a priority in their politics (a common response to slow, process-laden catastrophe).

In such situations, we need a very different communication strategy, and that is what the arts are good for. Schopenhauer is right in the sense that the arts can deliver what science cannot. CAE does not want to go too far here. We are not advocating art as the metaphysical creation of universal, eternal beauty (although in nature the appreciation of beauty can be a potential intercultural point between many environmentalists and those indifferent to environmental crisis, and therefore a place to begin dialogue). To the contrary, CAE has continuously fought against such principles as inherently authoritarian, and considers that element of Schopenhauer's thoughts on art and aesthetics as one to be avoided, however influential and institutionalized it may have been. As artists, we are on a twofold mission: one, to deal with difficult audiences—the nonbelievers, so to speak—and develop the tools and situations that make communication possible. Schopenhauer's value here is that he reminds us that at times we have to leave the charts and graphs at home, and find ways to communicate crisis, in particular, to audiences beyond those already convinced that climate change and mass extinction are taking place. What is the DDT for this century?[2] How do we find the intersecting symbolic that spreads the environmentalist perspective to an extent that it can be transformed into votes or policy in places where such activity is severely lacking? We will offer some suggestions in later chapters.

Finally, Schopenhauer is an early reminder of the ideological possibilities that accompany thoughts on nature in the face of evolution and evolutionary biology. Schopenhauer's secular and unflinching assessment of natural development is to be admired. As science tells us, the divine is nothing more than a wish, and if we follow the elements of evolutionary theory that Schopenhauer first proposed to the world, there is absolutely no reason to care about the environment, extinction, or biodiversity. And yet, through art and aesthetics, he gives us every reason to care, and to hold this contradiction as a necessary part of being (the ability of many environmentalists to hold this contradiction without cognitive dissonance is another of Schopenhauer's legacies). But we must go further, to either create a new artificial scale of value that need only offer a practical alternative to humanism, or support an already existing one. CAE is of course open to the former, but sees no evidence of its successful development, and tends to prefer the expanded contemporary version of humanism, because, like it our not, our biases and prejudices in regard to nature, which all too often go unacknowledged, will guide the policies that are made, and the choices of what is to live and what is to die.

Notes

1. Schopenhauer is a strange phenomenon. He stays in the philosophical canon, even though few outside of the arts have any time for him. He is slim fodder for dissertations or citations, but he has not found his way into obscurity (unlike Herbert Spencer, at one time the West's most influential thinker).

2. A few DDT-like symbolic alarms that have intensely reverberated in the popular imagination have sounded since Carson. The first was "the population bomb," from a book of the same name by Erich and Anne Ehrlich, written in 1968 and published by the Sierra Club and Ballantine Books. This neo-Malthusian treatise was alarmist to the core, and is yet another example of the savage necropolitics that emerges when environmentalism and alarmist demographic study bump up against one another. (The same problem can be found in Fairfield Osborn's 1948 neo-Malthusian book *Our Plundered Planet*, which was an influence on the Ehrlichs.) The Ehrlichs took their title from a 1954 pamphlet by General William H. Draper. CAE's concern is not with the book itself, but with how the phrase caught the public imagination. Why a general would like the metaphor of a bomb is obvious. The Ehrlichs, being more scholarly, originally preferred the title *Population, Resources, and Environment,* but understood the value of marketing, so "bomb" was used instead (with the blessing of Draper). Needless to say, bombs and explosions truly resonated in the minds of Americans in 1968, whether it was the baby boomers or the World War generation. Daily reminders of the bombings in Vietnam flooded the media, and Americans did not want to see this destruction come home. As the war wore on into the 1970s and the resistance to it intensified, the bombs did come home as the radical left began to blaze a trail of revolutionary adventurism with bombs used as calling cards. The "population bomb" remains a part of the language, and the anxiety it evokes continues among some demographics to this day.

As the population bomb/explosion began to fade from the public imagination, a second alarm sounded in the 1980s, with concerns over the "hole in the ozone." That sounded bad. The atmosphere was failing in its protective function and allowing the sun's dangerous, cancer causing, high-frequency ultraviolet rays to strike all of earthly life. This concern went so far as to lead to universal international cooperation to phase out a known cause—chlorofluorocarbons—in aerosol sprays and refrigerants, which appeared to be a good idea because it did help with slowing ozone depletion, and then with its rebuilding. Unfortunately (or fortunately, depending on how the problem is viewed), there was no "hole." What scientists meant was that they were observing a substantial depletion of ozone over Antarctica, which was not as aesthetically satisfying as the thought of a roving hole over populated areas zapping people with skin cancer. For those with a sense of black humor, the irony is that ozone is a greenhouse gas, and could contribute to melting Antarctica faster than it already is, which will in turn threaten many coastal and island communities. The jury is still out on whether the current rebuilding of the ozone to 1950s levels is good for the environment or not.

The third and most recent symbolic alarm is the Great Pacific Garbage Patch, which is usually described as a floating landfill the size of Texas, and at times much bigger—the size of North America! The idea of a mass of plastic bags, drink bottles, detergent flasks, and best of all, babies' and children's toys whirling about the Pacific choking the life out of it, seemed to implicate everyone in the ruination of the oceans. The island had to be eliminated or at least reduced. However, much like the ozone hole, there is no island, nor landfill. The Great Pacific Garbage Patch is actually billions of shredded, but mostly particalized, bits of plastic suspended in the seawater in multiple Pacific Ocean convergence zones, covering areas for which no one has an accurate scientific measurement. Not only can people not see the garbage patch from space, they might miss it if they sailed right through it. Without the image of the great patch, the pollutant lost its unbearable majesty. No international cleanup initiatives appeared despite all the mass media hullaballoo, but the (false) image did appear to reduce plastic usage among consumers intrigued by it. What these examples tell us is that threats to life on earth (particularly when they occur elsewhere or in the future) need to be aestheticized in order to carry moral or environmental outrage. We might also note that specific plans (the elimination of DDT or chlorofluorocarbons) function better than general plans (global population reduction or plastic-free oceans), much as environmental fixes for imminent event disasters occur more quickly than those for slow disasters like climate change.

Black Rhino

4

Necropolitics and Wildlife

One Lockean principle that most across the political spectrum can agree with is that the administration of death is a hallmark of society (even if it is not a hallmark in which we take pride). Locke thought of it in a very limited manner, primarily as a judicial act as punishment for an injustice. Unfortunately, numerous power constellations have means to decide who or what is to die, when, and for what reason(s), to transform these decisions into law or policy, and then muster the power to enforce them. Consequently, this chapter could go in many different directions to illustrate the scope and the depth of necropolitics. In truth, this should be a multivolume set, but we leave that for a more ambitious scholar, and limit ourselves to some timely illustrations.

Those Who Are About To Die

Before we get to the necropolitics of wildlife, we should start with a few illustrations in the human world. The military is an obvious yet arbitrary place to start. CAE suggests beginning with the 1960s political icon of necropolitics, Robert McNamara—and not just for his role in the quantification and escalation of the war in Vietnam, but for where he originally made his name in World War II. McNamara served from 1943 until 1945 with the Army Air Force Office of Statistical Control, where he administrated a bomber command of B-29s for which he developed schedules that allowed their dual use as bombers and as transports in India and China. However, McNamara's crowning achievement was that he and his team, slide rules in hand, devised the most effective and efficient (i.e., in terms of maximizing death) way to bomb Japan. Given that the cities of Japan were primarily made of wood, McNamara believed that firebombing would have the most devastating effect and maximize civilian casualties (missions aimed at killing noncombatants were called "morale" bombings). Guided by the bureau's calculations, the US Air Force burned to death upward of nine hundred thousand Japanese—one hundred thousand in Tokyo alone. For these war crimes, McNamara received a Legion of Merit (an Armed Forces medal given for "exceptionally meritorious conduct in the performance of outstanding services and achievements").

The battle of the calculator can go the other way. Today, North Korea is calculating how much death it must be able to produce in order to assure its sovereignty. The North Korean oligarchy (a small network of postwar first families) has now intergenerationally transformed North Korea into an impregnable fortress. They know that they are too weak and isolated ever to leave it, but they can make sure no one hostile ever gets in. Since the cease-fire, the immediate threat has been bombardment and surgical strikes, primarily from the US and regional rivals. They answered by moving military operations deep underground, and by deploying massive, well-protected artillery emplacements that could devastate Seoul. Whatever type of military assault comes their way, short of full invasion, they can immediately retaliate with an equal or greater deadly and destructive outcome. The hope is that the threat that fortress North Korea has created will escape a full-scale invasion until their nuclear program can be completed. Once it is, the calculations of death will make even a full-scale assault impossible because the violence could lead to global devastation. Although they cannot venture out of the castle, they can maintain their regime and their borders without outside interference, no matter how determined an adversary might be. North Korea is a fascinating case to those interested in the topic of necropolitics, in that its organization of life (biopolitics) is subservient to its grand necropolitical initiative.

Taking a step down from the military, the police are the next related category to consider. At present, ground zero of this necroadministration is the disproportionate killing of young black men, consistently demonstrating that "all lives matter" is at best poorly aspirational in the US. What disturbs CAE is the thought that protest may not be effective in curbing this unacceptable tendency in policing. We are convinced that it will help to ensure better treatment of protesters, that it may diminish racism in a general sense, and that it may even affect police policy and organization, but we are skeptical as to whether it will change the courts, which is where the problem confirms its perpetuation. As we have seen since Rodney King, the courts are immune to protest or riot. As long as an officer may offer the testimony "I *felt* my life was in danger," or "I feared for my life," and as long as such a phrase can function as magic words that excuse any act of violence, the problem will not stop, because the acts are consistently forgiven by the court and have modest social consequences for the shooter. CAE brings up this example because it is so different from the necropolitics we see in war. The lynchpin of this problem is that the subjective state of the perpetrator is allowed a legitimacy and an acceptance that is rare in necropolitics. Usually, necropolitics is administered through the calculating language of science, technology, engineering, and mathematics (STEM). As a means to distance administrators from the fact that subjects will die given certain calculations, forms of organization and acceptable behavioral tendencies are instituted. For example, those making the calculations or finding the "solutions" to problems are not expected to consider any controversy surrounding any decision. Under veiled threat, and present reward, they are excused from the politics of what they are doing. Moreover, subjects are eliminated from the process and represented as objects (usually in the form of numeric representation—see Appendix 1). That which can trigger emotion or aversion is minimized. But in the case of judicial review by jury, the opposite is at work; emotional response rather than "reasoned" assessment carries the day. No one knows how far the "feared for my life" defense can be taken. "I had to shoot all five people in the car because I felt my life was in danger." A subjective state cannot be disproven (feelings are facts), and testimony must be accepted on its face. Blue lives really do matter given this striking level of privilege.

Switching back to the STEM approach when examining racial policy in regard to necropolitics, CAE can point to the absence of environmental justice in areas primarily inhabited by black and brown people—meaning that all corporate externalizations (in this case, pollutants left for public cleanup) seem to make their way to landfills and remediation centers that are always conveniently located in poor, typically black or brown, neighborhoods. If we go to ground zero in the US, we would land in Louisiana, the most polluted state in the nation. The state has put all its financial hopes for its industrial base into one basket, and that is nonrenewable energy extraction and refinement. Given that Louisiana has no competing investment beyond tourism, the sole center of economic power—the extraction industry—has completely seized political power. It is either exempted from paying taxes, or pays at a very reduced rate. Perhaps more significantly, there is also no state enforcement of pollution standards. Only the overstressed federal Environmental Protection Agency can act, and they do not. Perhaps because it is a lost cause, perhaps because only the poor are suffering—but for whatever reason, the toxins continue to flow, particularly down the stretch of the Mississippi River between Baton Rouge and New Orleans now and probably forever known as "Cancer Alley." Along this eighty-five-mile corridor, town after town has died or is dying a slow death. People have no recourse, as the corporations own the state and local governments. The best outcome is to be bought out of one's home (as in Norco, Louisiana), and that is a terrible outcome. Imagine being forced from a family home, often after generations of living there, losing one's community, and doing so in poor health from years of exposure to an array of toxins. For others it is much worse—it is simply exposure and death. All calculated; all understood by the industry. Land is needed for the extraction industries' toxic processes. This land needs to be close to the base ingredients but as far as possible from those who might object with sustained and funded vigor. The point of intersection is usually where those who are most defenseless reside. Having control of the zoning laws, the industry can buy up property in the designated area and degrade resistance through the slow poisoning of the resident population.

To dovetail to another related necropolitical subject, do these unfortunate residents have access to healthcare to reinvigorate their poisoned bodies? Overwhelmingly, no, they do not. But as usual, whether it is with social or environmental justice, Louisiana is a beautiful test site. In 2010 when the Affordable Care Act was enacted, the Louisiana governor (a cancer in himself whose name CAE will not repeat) refused the federal dollars to expand Medicaid (health insurance for the poor). He was replaced in 2016 by Democrat John Edwards. Within two weeks of his inauguration, Edwards signed an executive order to accept federal money and expand Medicaid. The consequences were immediate:

On June 1, 2016, the expansion began, and within one year 378,564 people had enrolled in the program.

50,622 members received preventative care.

Nearly 5,000 women completed diagnostic breast imaging, such as mammograms, MRIs, and ultrasounds, and 63 were diagnosed with and treated for breast cancer.

2,276 patients were newly diagnosed with hypertension.

More than 11,500 new members received a flu shot.

4,474 Louisiana Medicaid enrollees had colonoscopies. Out of that number, 1,126 had precancerous polyps removed.

Yet this swing in the distribution of healthcare is probably not going to last if the conservatives in the US congress finally get their way. Attempts are being made to repeal the Affordable Care Act or at least stop the expansion of Medicaid. Given the first two bills proposed as replacements, according to the nonpartisan Congressional Budget Office (CBO), somewhere between twenty-two and twenty-four million people will lose their health coverage. For the grand majority of those losing healthcare, the enactment of this legislation is going to affect quality of life and life expectancy, but even more shocking is the CBO prediction that over the next ten years, 208,500 people will die due to lack of care. It is one thing to remove access to a "product" (which is how the conservatives think of healthcare), but the 208,500 figure represents the outright taking of lives, and cannot be described as anything other than that. Again, this is why God is so important in the necropolitics of conservatives in the US. Religion tempers the cynicism of needing to maintain power at any price, by allowing the belief that God wants to punish those who failed to lead a just and providential life (idlers), and reward those who have been loyal to His command (in this case with tax breaks) for they will fulfill the world's potential. (After all, the plutocrats are the opportunity and job creators.) Death, for conservatives, is a natural regulating power in the Malthusian sense, i.e., death by neglect (although they are every bit as willing to dole out death through active command, as in war or via the death penalty).

This idea of healthcare stands in sharp contrast with that of the American Medical Association (AMA). Not that the AMA has always walked a straight and shining path, but it has shown a level of ethical responsibility seldom seen in healthcare in the US in its support of the Affordable Care Act. As is to be expected, the AMA's view is quantitatively based, but perhaps this is one example where quantity is a better trajectory than quality. We can see the basic assumption in old yet ongoing medical processes like triage. Given finite time and supplies, prioritize services that save the most lives with the resources available. Single-payer healthcare systems are based on this same idea. While more unique treatments, both operational and pharmaceutical, may be under-resourced, preventive and standard care is available to all, which in turn increases average life span. For medical outliers, this system is less than ideal, as the lives of this marginal class of people are almost certainly shortened. This is the failed side of a necropolitics of utility in a system of limited resources; however, the alternative, in which medical treatment and medicine is distributed so that those who can pay get full coverage and those who can't get nothing, leaves even more to be desired. Given its class position, the AMA is taking the odd step of devoting itself to collective interest over individual interest.

CAE could go on and on with example after example of necropolitics at work, but we will instead stay on point. In Chapter 1, we touched upon the lack of an intentional necropolitics in the environmental movement, and we will now examine necropolitics as bureaucratized and articulated through the managed (intentional) death of wildlife.

The Nonrational Organization of Care for the Wild

Wildlife management, when distilled to its basics, often comes down to two general objectives: conditions leading to a low probability of population collapse or explosion should be maintained, and alien species should be kept out of the system. Accomplishing these objectives allows for the preservation of the delicate balance that is indicative of a healthy environmental gestalt. However, managers will be the first to say that nature (environment and wildlife) is always changing and evolving and with that comes a terrain of shifting equilibrium. In addition, species are introduced into and removed from the system (naturally, postnaturally, and artificially), and populations are artificially contracted or expanded on a regular basis. Management models used for the purpose of maintaining healthy environments have primarily emerged out of STEM culture. They seek to be objective and quantitatively based (and to a large degree they are), but as we shall see, aesthetics and ideology always find a way to insert themselves into both theory and practice, not to mention the nonrational desires and biases inserted through various public pressures.

From the point of view of maintaining a stable environment that is biodiverse (which, as we have seen, is an aesthetic and economic imperative), deer in the Northeastern US and Quebec, Canada, are counterproductive creatures in relation to these objectives. There are two primary reasons the deer population has run out of control: human determination to eradicate wolves in the region, and the fact that humans have unintentionally engineered rural land to be as perfect for deer as it is for humans. Considering that the uncontrolled herds are wiping out the plants that constitute their food source and allowing for the explosion of the few plants they will not eat (black cherry trees and hay-scented ferns are among the worst culprits), they can now be considered ecological pests. Unfortunately, "pest" is not the language of science even though it is used by biologists and conservation experts. "Pest" is ultimately an aesthetic term that describes whether a creature provokes a subjective state of annoyance in humans. A pest for one person is a charismatic animal for another, an important food resource for a third, and an object of sacrificial amusement for a fourth. An odd alliance of pressures coming from hunters (both food and trophy) on the one hand, and "save the deer" proponents on the other, is forcing wildlife managers to allow the deer to remain overpopulated. The public demand is that deer should be readily available whether a person wants to view them or kill them. Many of the "saving" campaigns do at least understand the dangers of having an exploding deer population and suggest nonlethal means for relocation, which is where they break with the hunters. The managers are happy to let the hunters harvest the deer, especially if they will shoot females. For some resource-oriented deer hunters this is a fine idea, but the rest of the hunters want a trophy buck, whose death does not help to reduce the population (unless the bucks are severely pressured and there are not enough to service all the does; unfortunately, when bucks become that rare, the effort to find them becomes too great and hunters tend to move on). What is left is an ongoing ecological problem that is not being solved and that makes a person wonder whether democracy and rational land and wildlife management can coexist.

A similar situation arose in the Galápagos Islands in the 1990s with feral goats. In the sixteenth century, the Galápagos Islands were seeded with goats and pigs by whalers and pirates who needed food caches. The islands were a convenient place for them to stop, and they also liked to eat the Galápagos giant tortoise. The tortoises were especially appealing because they could be stored alive for extended periods with little trouble. The goats and pigs added variation to the sailors' diet. As these practices disappeared, the need for the goats did as well, and the goat population exploded. On the island of Isabela, the goats were naturally penned in by a volcanic rock barrier of extraordinarily hostile terrain. Unfortunately, some goats made the crossing over the rock sometime in the 1970s. By the 1990s, this relatively small goat population had boomed into a destructive horde that threatened a broad variety of plant, insect, and animal populations all over the island. The goats had eaten the environment to a point of near desertification. For the most part, no one cared about most of these vulnerable life forms, but there was one that was so charismatic that it had to be saved—the aforementioned tortoise. In order to bring the devastated environment back to one that was suitable for tortoises, the goats had to be eradicated. If the goats were gone, biologists and conservationists correctly believed that the environment would heal itself. This was the policy that came out of the Tortoise Summit held in the UK in 1995 to address the goat problem. As was to be expected, there was some pushback on the idea—the goats had as much right to be there as the tortoises—but it was not enough pressure to sway the decision of the experts. The eradication program, Project Isabela, began in 1997. The goats were to be wiped out, not just on Isabela where they had reached pest status, but on all the islands as a preventive measure. This initiative seemed more than reasonable, as it would save a variety of rare and vulnerable tortoise species at the expense of an animal that, due to its usefulness to humans, was among the most successful mammals on earth.

One thing that can be said with certainty about those who inhabit STEM culture is that once they have a mandated, clear objective, nothing will distract these actors from their mission, and in the Galápagos their mission was to kill every single goat. Killing the first 90 percent was not difficult. Herds were rounded up by helicopter and shot on site. The final 10 percent was a bit more difficult. As the goats became rare, they became harder to find, so a great deal of effort was required to root out this small band of survivors. The search was an expensive undertaking, so a more efficient means for finding the goats had to be formulated. Since female goats will not herd when raising kids, the solution was to capture some female goats without kids, sterilize them, and artificially put them into extended heat for 180 days. (This, of course, is animal cruelty, but that was a distraction to be overlooked. Humane action is often a first casualty once eradication orders have been operationalized.) Next the goats were fitted with radio collars and released. Due to their social nature, these "Judas" goats (770 were used on Isabela) would find other goats, and vice versa, and the herds could then be tracked, rounded up, and shot. This was repeated every two weeks until the population dwindled to one herd made up entirely of Judas goats. In 2006, the eradication of all the islands was completed—approximately 250,000 goats were dead. CAE should also mention that all other introduced species were also eradicated, including pigs, donkeys, and rats.

While the eradication program was happening, another problem was brewing. The fishermen in the area were doing very well for themselves harvesting sea cucumbers, and did it to the point of overfishing, causing the park service to step in and put limits on sea cucumber harvesting. This lead to an uprising among the fisherman who felt their survival was threatened. The policy of putting sea cucumbers before people infuriated them, and they struck back with violent demonstrations, burning park service buildings, blockading roads—and, just to show they were serious that people should come first, they began slitting the throats of tortoises. To make matters even worse for the park service, the fishermen also started reintroducing goats to the islands again (a very subversive form of biological warfare). To counter this measure, the small herd of Judas goats (Isabela by then had 266) got to stay for monitoring and control purposes. The park service learned that policy by expert advice alone is probably not the best way to govern, and they did begin to integrate various communities and stakeholders into park and environmental decision-making processes. In the case of the fishermen, the uprising fizzled as many discovered they could make more money ferrying tourists around the islands with their fishing boats.

Parks from Africa to the Americas make necropolitical policy for the wildlife in the parks—everything from culling to eradication. Given that death is a fundamental part of management with the goal of sustainability, why not turn it into an asset? Park service officials who need to be paid for their service can do this grim work, thereby depleting the parks' budgets, or people who want to do it and are willing to pay for the privilege can do it, thereby growing the park's resources. For this reason, some parks grant licenses to hunters to come into the park and eliminate that which, according to park policy, needs to be eliminated. However, for those concerned with the humane treatment of animals, it is hunting that should be eliminated.

In 2014–15, this particular form of public pressure came to bear on Dallas-based hunter Corey Knowlton, who won a "conservation auction," giving him a license to kill an endangered black rhino (approximately five thousand are left in the world). The price he paid for this hunt was 350,000 USD, with the majority of the proceeds going to the Namibia Ministry of Environment and Tourism (which is also the institution that granted the license). The ministry had, at the time, identified seventeen other rhinos that needed to be put down. For the International Fund for Animal Welfare and the Humane Society International, this was unacceptable, as it was for many who object to hunting on its face. There was not much that either of these organizations could do to intervene in a hunt occurring in Nambia, but other protesters threatened the lives of Knowlton and his family, who required security services during the planning stages of the hunt. Knowlton did go through with the hunt, which was supervised by local professional hunters sanctioned by the government, and was filmed by CNN.

Knowlton, a professional hunter himself, believes that he has a stake in the continuation of the species, and that through hunting he can help—and not just because of the money that goes to the park to protect the rhinos' habitat and the rhino itself from poachers. The black rhinos tagged by the ministry for hunting were old and could no longer contribute to the gene pool, but they could subtract from it. These rhinos are fiercely territorial, and will not hesitate to kill a younger rhino unlucky enough to cross their path. Conservationists see this need for hunting as a necessity, especially in the case of an endangered species. Old rhinos who are out of the gene pool cannot be allowed to kill those who can grow the species. Letting old black rhinos live will lead to an overall net loss to the species. A more humane means could be used to dispatch the animals, but at the loss of license money (that ecotourism, however, could potentially make up for). In the end, there is no rational way to solve the conflict between fundraising conservation hunting and anti-hunting groups. It will be decided by emotion, aesthetics, and regional norms and traditions.

These colliding moments between the rational and the nonrational are probably most representative of how a less formal necropolitics becomes policy. As long as biodiversity remains essentially an aesthetic choice, this should continue to be the case. In 1995, wolves were reintroduced into Yellowstone National Park. Reintroducing a predator to control grazers and browsers that if left unchecked can be quite destructive was a rational policy. But there was also a nonrational reason. Since we are no longer agrarians, we like the idea of wolf packs roving the park. They are an awe-inspiring part of the scenery. We see them as free, strong, and noble. The wolf truly seems to inspire this projection along with empathy that comes from this very anthropomorphism. This is an all-too-human mix of reason and desire, which we can find as far back as the writings of one of the first systemic conservationists, Aldo Leopold, who described his thoughts and feelings about wolves in this now-famous quote from his book *A Sand County Almanac* (1949):

> We reached the old wolf in time to watch a fierce green fire dying in her eyes. . . . I was young then, and full of trigger-itch; I thought that because fewer wolves meant more deer, that no wolves would mean hunters' paradise. But after seeing the green fire die, I sensed that neither the wolf nor the mountain agreed with such a view.

The Evolving Organization of Care for the Wild

Biologist, ecologist, and wildlife manager Allan Savory provides a very interesting model of a never-say-die approach to the dying. It took nearly a lifetime, but he may have found a solution to the problem of desertification—green ecosystems being transformed into desert, much as we saw in the Galápagos example. The common wisdom is that the primary cause is overgrazing by domesticated animals like cattle, sheep, and goats. Growing up in southern Africa, Savory "loved the land and wildlife more than [he] hated livestock." As a game officer in Rhodesia (now Zimbabwe) in the 1950s, there was not much he could do about the entrenched ranching industry, but he could get livestock out of the parks and protected areas in the hopes that their absence would stop desertification. When the desertification process did not stop, the blame fell to one of the great wild grazers: the elephant, which was thought to be not only overgrazing but also stomping the ground to death. Savory and a team of government-sanctioned experts believed the best way to stop the desertification process in the park was to shoot elephants to get the population down to a level that would sustain the land. This policy was enacted, approximately forty thousand elephants were exterminated, and desertification still did not stop. Savory later said, "That was the saddest and greatest blunder of my life."

This is the problem we began with in the introduction. Ecological systems are so complex that scientists and wildlife managers are often working with little more than educated guesses combined with trial and error, in the hope that what appears to be the obvious answer is in fact the right answer in managing land and wildlife. In this case, killing wild grazers was not the right answer to desertification when massive tracts of land are involved. The companion policy for stopping desertification is the burning of grasslands in order to rejuvenate them. This leaves the soil bare, releases the carbon held within it, and releases pollutants into the air. Worldwide, billions of acres go up in flames. Needless to say, this solution is also rather undesirable. Leaving these relative failures behind,

Savory's new hypothesis is to mimic nature through what he calls "holistic planned grazing," which involves reestablishing large herds in vast grassland areas combined with simulated predatory attacks that keep the animals moving. With this type of "naturalized" movement, the grasses will be fertilized, mulched, and stamped enough to not only rejuvenate, but also hold water and carbon in the soil and break down methane. In test areas of fifteen million hectares on five continents, the mimicking process, according to Savory, seems to work in reversing desertification. Savory goes on to claim that if all the grasslands can be restored, enough carbon can be taken out of the atmosphere to return to preindustrial levels, ending human-created climate change. This all remains to be seen, but this model is an interesting assemblage.

The punchline here, however, is where to get the animals for this massive undertaking. There are not enough wild grazers left, but there are the domesticated grazers who are among the most successful species on earth: cattle, goats, and sheep. And there are plenty on every continent where they are needed. Love your cows and get ready for the price of meat to come down and for protein-rich diets to become more common among those with few economic resources. Of course, we have been told for decades that one of the best ways to contribute to a reduction in climate change and maintain biodiversity is to become vegetarian. And those forces are mustering with furious objections to Savory's claims. The counterforces state that Savory's experiments cannot be replicated, nor can his mimicking strategy scale under most grassland conditions. The opposition can be summed up in the punchy slogan "There are no beef-eating environmentalists." This may be true. We have no idea if such a plan could work, but CAE still likes Savory because of *how* he is thinking. He is not saying that we must eliminate humans, but asking how we fix the environmental crisis with the materials and know-how we currently have. He is working on a different kind of assemblage to revive multiple parts of the environment. This is a model for how problem-solving around the environment should occur. Unfortunately, CAE believes that this debate will be argued on an emotional basis, but it could eventually be solved on a rational basis through data analysis combined with economic pressures. Unlike with deer populations, aesthetics will have a smaller role.

While in this section on the organization of care through trial-and-error initiatives, CAE would like to comment on art and environmental care. Recently, the nonprofit initiative Robots in Service of the Environment (RSE) has attempted to solve the lionfish problem in the Caribbean and the Atlantic.

By means unknown, the lionfish has found its way into these waters. It is quite adept at reproducing and has no known predators in these oceans. Consequently, the population is exploding. The fish is edible and does appear on some menus, but is never in consistent enough supply to really establish a niche in the seafood industries. The fish does not school, so it is not ideal for commercial net fishing, and is mostly killed by divers spearfishing. RSE has created a robot that can swim up to the lionfish (who, owing to its lack of predators and its great defense system of poison quills, is not afraid of other creatures, natural or artificial); the robot then shocks it, and sucks it up into a container that can hold up to ten fish. Clearly, the lionfish catcher is just a proof-of-concept machine. It cannot be used commercially. Catching ten fish at a time is not going disrupt the expansion of the lionfish population, even with a thousand of the robots on patrol.

However, perhaps we are looking at this machine in the wrong way. CAE believes we should see it as robotic art. The robot is about potential, and holistic thinking—suggesting a possible symbiotic relationship between engineers, the fishing industry (perhaps amateur), marketers, and restaurateurs. The project is about making new assemblages that can suggest a path to a better environment. It is what the artist Tania Bruguera calls *arte útil*, or useful art—tools and devices made by creative people who rearrange intersectional disciplines to make them do what they normally do not. While this project may not be a triumph in engineering or economy, it is a very solid glimpse of what the area of robotic art and the environment might look like in this century. This project is of particular interest because it appears to be the first, or among the first, examples of positive necropolitical robotic art.

The Rational Organization of Care for the Wild

We live in the Anthropocene. Wilderness is no more, but generally speaking, in the human-occupied zone there has not been much wilderness for decades. Lands often thought of as wilderness have been managed using techniques such as control (reduction or removal of species), harvesting (hunting and fishing), introduction of species and pathogens, controlled burns, and barriers and fencing. Some or all have been used worldwide in an effort to bring a rational method of care to wilderness, rendering it no longer wilderness. Just as wilderness no longer exists, neither does a system of wildlife management that enjoys scientific or public consensus. What does exist are various models popular among wildlife professionals—that, as with most disciplines, constantly mutate with theoretical fashion. When these models are applied, we cannot be sure how well they are going to function at a material level, and every wildlife bureaucrat knows this. Moreover, as we have mentioned with regard to STEM culture, even when those involved recognize controversy (usually political in nature), they do not account for it in their models, which is yet another reason why the functionality of a policy once it hits the ground is unknown and can bring surprisingly unexpected results. What biologists and wildlife managers are expressing in their models is *what* they would do, and *why* they would do it in a certain way, in order to meet a given environmental objective that at root is actually political and cultural.

CAE believes there are three basic theoretical perspectives in regard to wildlife management, each of which can be broken down into more nuanced theories. The first and oldest perspective is scenic preservation (especially as it applies to parks), which includes the preservation of land and charismatic animals. This perspective is mostly aimed at managing human interaction with the land, which is usually the most destructive factor. It seeks to control the quantity of humans accessing the land, what they may do on the land (e.g., look as opposed to impact), and how the space is developed (if at all). Development would include roads, trails, campsites, lodges, water delivery and distribution, etc. This objective, the first of its kind, is no longer the most significant priority; it is still a major factor, but other elements have become of greater concern.

The second perspective is based in support of biological diversity. This can mean either species diversity or habitat diversity (usually the latter). This perspective would also include those who frame the primary objective of management as "genetic variability." The professionals advocating for this directive are very prone to intervention to maintain or improve the environment relative to the objective. The problem is that protection parameters are set in accordance with what are believed to be indigenous species. On face value this sounds good, but as we know, ecosystems, like the members in them, change over time, so what is the temporal designation that marks the status quo? Should it be when humans first inhabited the area? When white people first saw it? When those of a scientific persuasion started to catalog it? Or the way it exists the day management begins? This of course then spills into the consideration of the traditional cultural norms associated with the area, and what people should be allowed to do and when. Fortunately, these questions are always some other agency's problems to sort out. Until then, they can be ignored, and an arbitrary date will be assigned from which management will proceed.

The final perspective, and this is the most recent, advocates for a more systemic and process-based approach. A complex set of evolutionary, physical, genetic, behavioral, and ecological processes should be maintained in conjunction with compatible or co-evolved populations. Consequently, this approach is a little more open-ended in terms of what belongs in the system, and does not seek to fight change in habitats or communities as long as the processes remain healthy. This is an interesting yet paradoxical model in that it is compelled by wildness, but also by intervention. It seeks to create an assisted form of evolution that understands that natural and cultural processes cannot be disentangled. "Compatibility" appears to be the first operative term for loosening the interpretive possibilities for this prime objective. The second key term that loosens the model is "process." While the first two perspectives discussed above focus on "states," this one accepts the results the processes yield until results outside the parameters of "acceptability" are predicted to appear (population collapse or explosion). For example, the snowy owl has been moving farther south in its winter migration and often staying south for longer before returning to the tundra. The prey populations do not seem to be suffering, and the owl population, while growing, has not achieved pest status. Its impact on the overall environmental processes is negligible. So while the owl may not be indigenous to these southern areas (at times as far south as Washington, DC), no control policy is necessary. On the other hand, necrointervention has been activated for the nonindigenous Burmese python in the Florida Everglades, because it is decimating the indigenous marsh rabbit.

Given these perspectives, how is necropolitical policy expressed in terms of rational analysis? One key problem that wildlife managers have to address is extinction. Populations go extinct for two primary reasons. The first is driven extinction. Some fundamental change in the environment drives a population down and finally to extinction. These drivers are powerful, so even as the decrease in population increases food resources and possibly lowers predation where applicable, the population continues to collapse. Short of an obvious environmental catastrophe, or a generally known problem like habitat fragmentation (and its spinoffs), these drivers are difficult to identify.

The second type is stochastic extinction—when chance events have terminal consequences. This tends to affect small populations. Because the populations affected are small, numerous variations and fluctuations that would not bother a large population come into play, making the identification of the destructive variable(s) extremely difficult. Moreover, this random form may occur via demographic or genetic malfunction. Causes could be anything from skewed sex ratios or age demographics to chance emergence of recessive genes that reduce a creature's fitness.

What is the solution to these types of extinction? Essentially, experts who study the creature and/or local habitat in question attempt to establish a comprehensive list of possibilities as to why the collapse is happening. Possibilities might include introduced predators, lack of food, unregulated commercial and recreational hunting, environmental contaminates, competition from introduced species, or introduced diseases. Hypotheses are prioritized and experiments are designed to test these hypotheses. If a factor is identified, and done so in time, corrective measures are taken.

In the case of a population close to collapse, rescue and recovery is the tactic used. Endangered creatures are captured and bred in captivity. Some creatures are amenable to this activity, while others are not. For those who are, once the population is large enough, approximately twenty or so creatures can be released as a test to see if the cause of extinction was properly identified, or if not identified, to see if the driver is still at work. The released creatures are a trial-and-error suicide squad. The likelihood is that the variable has not been identified, so everyone involved is simply hoping for the best as the radio-banded creatures are sent back into the wild. In spite of all the spectacle of science and the good intentions behind it, if the problem is not obvious, extinctions are often too complicated to solve. Populations undergoing stochastic extinctions are in heaps of trouble.

More typical of wildlife interventions are control initiatives. One form is a tactical response to a temporary problem in a system, such as too many feral cats or raccoons in a suburban area. They can be captured and relocated or killed. The event is short-lived, and the environment is returned to a stable state where the creature is downgraded from a pest designation to being an acceptable part of the ecological community. The second form of control is much more serious, in that it is aimed at moving a stable system into a different one that is considered more desirable, and yet is often more unstable. This type of action can easily become a powder keg when charismatic animals are targeted, although not necessarily so (for example, the control policy for deer in New Zealand is noncontroversial). With most cases involving noncharismatic animals, there are few complaints (as with the control of potential Zika virus–carrying mosquitos through the use of pesticides and control techniques such as manipulating fertility with genetically engineered males). The same could be said of other rodents or insects requiring control programs. These control initiatives are strategic plans that will continue from year to year in order to maintain the desired environment. Creature density is lowered, and the new density is enforced so long as the desired effect is achieved. The reduction of creature density is the means to an ecological end, and not an end in itself. Reducing the negative effects of the offending creature on others or to the landscape itself is the goal of control and is also how success is measured. This is what distinguishes control initiatives from sustained-yield harvesting (hunting as described in the example of the deer). The former is a management action while the latter is an objective.

Many nonlethal forms of control are available for use. These include fertility manipulation, genetic engineering (usually to disrupt the reproductive cycle), immunocontraception, fencing and barriers, alarming, and even food intervention (to keep an animal from eating a food source that needs protecting, managers can distribute an accessible food source that the animal prefers to the original vulnerable source). These forms of control comply with the imperative that the use of control be humane. In light of this imperative, the problem of the humane stirs itself whenever we talk about control through direct killing, along with its cousin sustained-yield harvesting. Here, the traditional method is still used: death by firearms (or bows for some hunters). Whatever method is used, given the complexity of attempting to move from one stable system to another—in which the original is overwhelmingly likely to be the more stable of the two—mistakes are going to be made. The desired outcome may not occur (as in the elephant example), meaning the designated pest animal is not actually a pest, or the control has an unintended effect on nontarget species. The one optimistic takeaway is that professional controllers know when they do not know, when they are making an educated guess, and usually, rather soon into the process, whether it is having the desired effect.

Before concluding, CAE would like to look to the near future and mention the development of a new technology that could radically change the necropolitics of control in the realm of environmental management and its relationship to public health. Scientists and engineers from all over the greater academic universe of biology are attempting to harness the power of extinction. They have found a way, at least theoretically, that a species can be totally eliminated through engineered gene drives. These synthetic gene drives allow for a desired gene to spread through a species that sexually reproduces, by radically skewing the probability that it will be passed on to the next generation. In the wild, naturally occurring gene drives tend to increase fitness, but in the lab they can be made to decrease fitness and even bring about species elimination, and do so in relatively few generations. The production of these synthetic forms of gene drives has been hyper-accelerated by the creation of CRISPR gene-editing technology.

Consequently, this development has caught the attention of the US military. The military is interested in knowing how gene drives could be weaponized and what, if any, defensive measures could be taken against the use of such biological weapons. DARPA is throwing a considerable amount of money at these two problems. The former problem speaks for itself. If a set of genetically modified creatures capable of changing the genetic makeup of a species to the point of causing eradication does exist, one of its functions is as a weapon. The latter is the problem still in need of investigation. DARPA wants to know what can be done in the case of bioterror (when fit creatures with gene-drive modification are released into the environment by a hostile power). Just as important, what can be done in the case of bio-error (if this weapon escapes the laboratory, how can a genetic spill be contained or reversed)? Unfortunately for DARPA, and they should know from past forays into bioweaponry, biological agents never prove sufficiently predictable.[1]

The other players are a small number of foundations concerned with global public health. The most significant is the Gates Foundation, which has already invested tens of millions of dollars into the research. In relation to gene drives, the foundation is interested in one thing: how to stop mosquito-borne diseases, with a big emphasis on malaria. The hope is that this technology will eliminate the various species of mosquito that carry the diseases. The Gates Foundation wants to save millions of people[2] and, of course, to keep from having to replicate the more than 39 billion USD already spent by the foundation since 2000 on trying to reduce, by various means, the harm malaria causes. They believe that the human death toll provides a very compelling reason to throw caution to the wind and move development and trials along at maximum speed. Of the more than 3,500 species of mosquito, around thirty will need to be eliminated, but the foundation and its allied laboratories are focusing on *Anopheles gambiae* (the primary carrier of malaria in sub-Saharan Africa). This represents a major necropolitical gamble, in that no one knows what will happen in various regional and local ecosystems when this species is eradicated—yet we do know what will happen to poor human populations in Africa if malaria is allowed to continue on its present trajectory. To complicate matters further, there is no rational way to make a decision about this matter. Science cannot help us here.

From an anthropocentric perspective, this may be a bet worth taking, but if ecological catastrophe follows it will not have been a smart thing to do from any perspective. For the sake of argument, let's assume the best—then what? If intentional extinction becomes an accepted mechanism of control, what might happen given present-day aesthetic and economic prejudices? Had this technology been available a century ago in the US, CAE is fairly certain that no wolves, big cats, foxes, or other predators that attack livestock would be left today. This technology expands the use of control beyond the limits of populations to include species. Moreover, it transforms the use of control as a conservationist strategy or tactic aimed at a specific locale into a global action (although there are conditions, such as on remote islands, where it might be used in a specific manner). A brave new world could be coming.

In the case of wildlife, we have the capacity to speak about necropolitics in a reasoned way, and among professionals this does happen. Yet even within this bubble, an underlying troubling consensus also exists that so little is known about the complexity of ecological systems that it is difficult to know what to do. Moreover, once these experimental forays of trial and error reach stakeholders—many of whom do not share a belief in scientific models for economic, political, or moral reasons—the process becomes even more unmanageable as more purified ideological forces work to shape objectives. Even in this nonhuman realm of wildlife management, where necropolitics should be the least conflicted, no common language is shared, and even the most rational scientific process cannot be purged of ideologically based pressures, choices, and biases.

Notes

1. Gene-drive technology may not work in the wild. If we continue with the example of the eradication of specific species of mosquitoes, the first problem is producing mosquitoes that are fit enough to compete and reproduce in the wild. Considerable effort is being applied to this problem, and it appears probable that it will be solved. The other, larger, problem will be natural resistance. DNA configurations that confound the gene drive might naturally emerge.

2. According to UNICEF, approximately one million people die each year, and over three hundred million suffer from the disease.

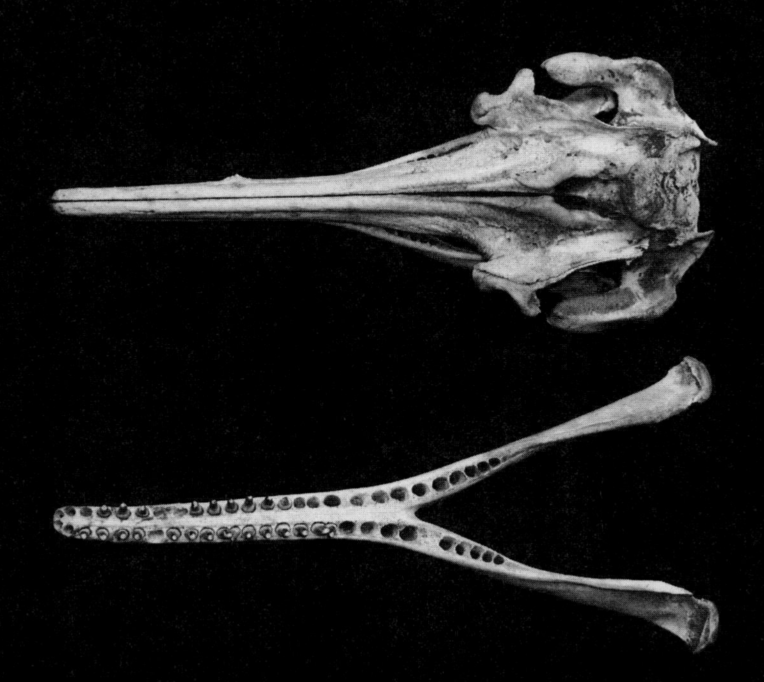

Amazon River Dolphin

5

Tactics: Reinventing Precarity*

Precarity has once again become a privileged category in economically depressed postindustrial economies, frequently used to describe everything from individual existence to the general social condition. Whether in domestic or public life, no one escapes the probability that a radical disruption of personal security or routinized social conditions could occur at any moment. While precarity is no more representative of life now than it has been at other moments in history or on other present-day points on the globe (in fact it is probably less so), it has become a noisier part of the collective consciousness as traditionally secure economic and ethnic groups make it a category that must be engaged before it is totalized as an irreversible narrative of the fear-mongering security state.

CAE considers here one manner through which precarity might be reinvented as a tactical form of productivity that resists the destructive initiatives of global capitalism in both rural and urban areas and in the environment in general. We explore the possibility of applying precarity's positive qualities in a manner that makes it more conjunctive with environmental

*A version of this chapter was originally published in *TDR: The Drama Review*.

struggle and environmental justice. The power of precarity in conjunction with law can, ironically enough, work in favor of a healthy environment. CAE suggests that we integrate the shared precarity of endangered plant species and endangered social and green spaces in a manner that strengthens and protects both. To initiate our *New Alliances* initiative, CAE traveled to Turin, Italy, in October 2011 to conduct a workshop on "new alliances" in collaboration with Parco Arte Vivente (PAV).

While CAE would not dare attempt to list all the tendencies that produce precarity in the postindustrial economy in a document as brief as this one, we will mention a few we believe are key. First and foremost is the reorganization of labor in a manner that allows various institutions of production to acquire every possible cent that social efficiency can yield. The intensification of administrative digitality, in association with improvements in real-space transportation, enables transnational capital to find and exploit the lowest labor cost and establish this bottom line as the basis by which all other nonspecific labor is measured. Consequently, the value of labor pools is very volatile. One can live in a company town (or city) one minute and a ghost town the next. A labor pool implosion causes a devastating cascade effect that sucks small business and local service industries down with it.

At the top of the labor ladder, the situation is a little better. There the workers have a specific desired skill (the digital equivalent of a craft or professional service skill) that cannot be replaced in a manner similar to replacing a standardized bolt in a machine. Beyond their "craft" abilities, these workers typically offer a diverse set of problem-solving and creative skills that can also be exploited. Moreover, they are completely flexible. They can work wherever they are, and they can work at any time for the precise amount of time a project requires. In order to do this, they have their own workstations, and assume both the costs of their training and the perpetual retraining needed to keep pace with the rapid change of the digital world. Consequently, in contrast to those at the bottom of the labor spectrum, they do bring some leverage to the table when selling their labor on the market. The life of these workers is feast or famine. They enjoy comfort and benefits while working, but must always be preparing for the in-between. The cost of acquiring the skills to reach this upper rung of the labor ladder is considerable, often amounting to the equivalent of a first mortgage and thus increasing the pressure for steady employment. Moreover, these local workers are competing with each other on a global scale, making for regional divisions that add to their particular type of precarity.

A second major tendency adding to the precarity of our age is finance capital's love for risk. Given that high-risk investments are the best way to maximize profit (if one is successful), high-stakes gambling in the financial sector is attractive to many. Because of this tendency, even the very wealthy are involved in the relative general condition of precarity. In the postwar US, when C. Wright Mills was writing *The Power Elite* (1956), the elite class was quite stable, consisting of family units of intergenerational wealth gleaned from manufacture, agriculture, or the extraction industries. Now, a sizable percentage of the elite class is part of the extremely volatile gambling class of finance capital. Members of this group can find themselves swimming in billions of dollars one moment and then crashing the next, left with only millions (or even landing flat broke). Yet as bubbles burst and legitimized Ponzi schemes collapse, there are not only the losers of the investment class, but the millions of small investors who are in the dark about what is actually happening to their investments. Retirement accounts are wiped out, homes are foreclosed upon, and the reality of downward mobility hits domestic space with full force. Profits may not trickle down the class ladder, but the material consequences of risk always make their way to the bottom—and not as a trickle, but as a deluge.

Within the social sphere, there is not much left to plunder but the public sector. In the United States, funds previously used for the public safety net are being handed over to elites in the form of tax breaks, corporate subsidies, and bailouts—the legitimized raiding of public coffers. And capital has no better way to extract money from the working classes than through war. The exorbitant military budget and de facto privatization of the military is a means to redistribute funds to the wealthiest through weapons manufacture and security services.

Finally, CAE must acknowledge the fundamental structural shift occurring in economies that are less dependent on manufacturing. As the economy moves from a mode in which industry dominates to one in which service becomes hegemonic, populations that have no place in the new economy begin to form. If no institutions exist to retrain those lost in the shuffle, these excess populations, the underclasses marked by permanent precarity, expand radically. Given that education is among the first casualties of austerity policies, the underclass will continue to grow; and not surprisingly, little is being done to ameliorate this situation. To make matters worse, many of the expanding areas of production do not require significant amounts of labor.

As global capital slips deeper into structural crisis and desperately seeks to maintain profit margins, becoming-precarious in a negative sense emerges as a dominant narrative. A heavy miasma of nostalgia for the stability of the past hangs in the air. Yet even if we could bring back the 1950s, who would want to? In the US, social improvement was linked to the intensification of accumulation, but for whom? Large marginal classes were not included in the enrichment of the social sphere by a capitalist system caught up in a "class compromise." And even many of those who *were* included did not enjoy the most desirable conditions. Men in "gray flannel suits" crowded the streets going to stable lifelong jobs devoid of satisfaction, doing what they had to and being careful not to rock the boat. Do we really want an encore performance from "organization man?" Do we want to go back to the hegemony of a family structure representative of the way only a small fraction of families actually live? Do we want to trade precarity for alienation and marginalization—a marginalization so profound that it will only further catalyze the ongoing cultural and political alienation of women, LGBTQs, and a rainbow of minorities? The answer instead may be to inspire a precarity that serves people and improves the social sphere, until it becomes possible for us to eliminate its negative aspects that function as basic conditions of life.

Dérive Revisited

Dérive (drift) can be interpreted as a utopian process. To be sure, it has great potential as a positive process for reaching a desired outcome. The drifter can break the routines demanded by normative structures and the dynamics of the urban environment. The drifter can resist the rational and let submerged desires that are stored in the unconscious guide the way. A drift should be unproductive, should lack practical performativity, and as such, will become adventurous. Drifters will mix with estranged environments and mingle with humans who exist on the fringes of their everyday existence. In so doing, experience is restructured outside the imperatives of the status quo. Drift requires active, engaged participation in immediate real space. Unlike strolling, it is not a distanced form of cool observation and data collection. Drift is rather a temporary demonstration of what liberated being-in-the-world could be if the disciplinary apparatus of the spectacle and the illusions of virtuality were not ubiquitous components of our lives.

When drift is described in this manner, the process sounds so pleasurable, and this pleasure seems readily accessible if only we would enact it.

However, there is another integral part of the process—a component that is generally glossed over, but always implied—and that is precarity. A drift could afford us a marvelous introduction to subaltern cultures we never knew existed, or lead us magically across the limit lines of a gated community of the 1 percent. Yet it could just as easily end in jail time or hospitalization. Surely if we gender or racialize this process or view it through the prism of unjust majority/minority relations of any kind, precarity intensifies. How will any agent charged with the enforcement of the status quo view someone in the midst of a drift? Jail would seem to be the probable outcome if the status quo of the social sphere is in any way perceived to be jeopardized. Certainly the Situationists, who first brought us this activity, were not so naïve as to think this way of acting would necessarily be wholly positive. Their documented physical risk-taking and their time in jail speak to their direct experience of the less-than-pleasurable aspects of creating and engaging in liberated moments and spaces. By proposing drift as a utopian process, CAE is not claiming it is without risk to one's bodily and personal autonomy. Drift rather seeks an intersubjectivity that authority forecloses because *to drift* suggests that a situation, or even the entire social world, could be other than it is. That is the utopian gesture, and its constant companion is precarity.

Precarity and Resistant Cultural Practices

The Situationists were not the first cultural activists, nor will they be the last, to count precarity as a constant companion. Any cultural worker who has performed resistance in a public space is quite familiar with this relationship. (As an operational definition, by "public" CAE means any space outside of the domestic that is not secured for specific access.) Breaking the law is often acceptable to authorities, because laws ruling comportment in public space are created less for stopping criminal activity and more for stopping resistant activity. For example, in spring 2010, CAE witnessed NYU students performing an action at the American Museum of Natural History in New York City that questioned why the institution would have a statue at its entrance of an equine-mounted Teddy Roosevelt leading a Native American and an African American (on foot, of course) into the implied sunset. The guerrilla performance primarily consisted of a dialogue between students who performed the statuary roles of Roosevelt, a freed slave, and a Native American. The discussion was as anachronistic as the sculpture, consisting of an attack on and an apology for "the white man's burden." As might be expected, security guards responded immediately, and police shortly thereafter. The police threatened the participants with arrest for blocking public pathways (in spite of the fact they were in constant motion). When students asked the police about the dozens of other people on the steps of the museum who were stationary and blocked clear access to the entrance, the police replied that *they* could break the law because *they* were not bothering anyone.

This is why CAE does not make an operational distinction between a park and a mall. The security is the same, and the demand of public order and limited speech is the same. Unfortunately, when resistant politics enters the realm of cultural production, a higher degree of risk has to be embraced. The laws to "maintain public order" are plentiful and function at differing levels of intensity. When an activity challenges—explicitly or implicitly—the status quo and the authorities that benefit from the status quo, laws governing public nuisance, public disturbance, disorderly conduct, unlawful assembly, or blocking public access are brought into play. If someone needs to be removed from the public sphere for a longer period of time, more serious charges are used, such as inciting a riot, causing a false public emergency, or criminal mischief. For authorities, these laws are helpful because they can be applied in a completely arbitrary manner. Anybody can be arrested at any time, and the arrest can always be framed as a deterrent to criminality rather than a means of quashing resistant performativities and minority representation.

Even when interventionists have the institutional cover of a legitimized sponsor to avoid legal troubles, other disciplinary agencies are waiting in the wings. For this situation, agents of the status quo, from politicians and lawyers to church groups and social workers, can take up the disciplinary slack. In such cases, cultural activists do not have to worry about jail, but the pushback can still be at best tiresome and at worst costly. Expression management is a ubiquitous phenomenon that permeates almost all of everyday life. As the structural meltdown of global capitalism flows deeper into crisis, resistant expression will be increasingly suppressed. Under such conditions, the cultural worker's relation to precarity will intensify.

Cultural activists have always functioned in a double bind in relation to precarity. On the one hand, creative cultural work tends to be economically undervalued for the grand majority of those who participate in this kind of production; however, economic impoverishment is often perceived as a fair exchange for a culturally rich, diverse, and even happier life. On the other hand, if suppression of activism is intensifying, precarity beyond the economic front increasingly becomes a part of social life for resistant cultural workers. We have no choice but to make it a friend by finding better ways to use it in struggles against oppression and social injustice. As a matter of personal testimony, at one time or another, CAE has faced almost every disciplinary agency imaginable, and yet we would not give up our relationship to precarity. The many empowering and pleasurable experiences have certainly outweighed the horrible ones.

Ecological Precarity

Perhaps cultural workers suffer from a third form of precarity beyond their precarious positions within the economic and political systems: ecological precarity, which is the general condition of existence for humans and many other species brought about by the purposeful, ethical bankruptcy of neoliberalism when it comes to environmental policy. Basing all economic activity around the principle of greed has created the most environmentally destructive conditions in history. Not only does capital seek to avoid any relation to the reproduction of the social beyond making sure the labor pool does not dry up, it has no relation to the preservation of life in any form. The biosphere is understood only as a resource to be used until depleted.

To be fair to the early capitalists, they could not have foreseen the current scale of the economy and how it would change the perception of the earth from that of a seemingly endless repository of resources to one where the end of resources is a near-future certainty. Yet once this fact was understood, why haven't capitalists veered from their ecocidal path? They have not because of their firm belief that nothing more than utilitarian value exists beyond an individual's ego. All objects (and that includes humans) are there only to be used and organized in a manner that yields power, wealth, and prestige. So long as the individual ego is not damaged in the process, the process is good. Any external death and/or devastation resulting from economic practices is simply a necessary sacrifice for the immediate glory of economic gain. Since neoliberals recognize the ego as finite, and its death as marking a real end of economic activity, they recognize no relation to the material world beyond this end. The future extinction of humans and other species is not recognized in their business plan. They are as removed from the infinitude of time as they are from the web of life. Now, within the ecosphere, capitalists are committing the greatest crime of their two-century history. From an ecological perspective, the capitalists have become a true death cult, in their belief that the weight of one of their own egos is greater than the weight of the universe.

The dominance of this ideology of perverse individualism, where the worst of human qualities are believed to be able to transform the world into a better place for all, justifies an ecological dynamic that causes everyone to suffer in the present moment from environmental devastation. At the heart of this matter are nondegradable and nonremediable pollutants. The worst assaults on the ecosystem come from this source. (CAE would be remiss if we did not note the strong second-place finish of the development and extraction industries for their serious contribution to environmental devastation.) The neoliberal perspective is that the cost of pollutants should be "externalized" (paid for by someone other than the corporation, which usually means the public). For the most part, the corporations have been successful in doing just that, to such an extent that the price of remediation is beyond the capacity of world markets. Whether we are talking about climate change, loss of biodiversity through mass extinction, decline in the quality of air, soil, and water, or health crises brought about by pollutants in the environment, the cost to fix what the earth in the Anthropocene cannot is incalculable.

Despite the neoliberals' worldwide success in polluting without legal consequence, they continuously work for even less environmental regulation. In general, neoliberals tend to dislike regulation of their own activities; regulating and managing everything else is fine, however, particularly if it involves any kind of market force like labor organizations, labor migration, competitive upstart organizations, or resistant activity. Then the government's job is to pass legislation that protects property and keeps "public order," and to enforce these laws and regulations to the letter.

Those who are concerned with environmental integrity have achieved some victories against the deregulation movement in spite of its massive wealth and (corrupt) political influence. Of particular importance now is the protection of endangered species and habitats in so many countries. In the US in 1973, President Nixon signed the Endangered Species Act (ESA) into law. He was, of course, forced into this position by environmentalists and a variety of concerned citizens, but also by some unlikely conservative allies in the grip of renewed fears about the consequences of overpopulation. This piece of legislation gave near-extinct creatures a recognized legal status, and this status gave environmental advocates a way to push back in the courts against those who were prepared to kill any living thing that stood in the way of profit.

New Alliances

Given the dual nature of species that are both endangered and protected, it appears we have a possible site where the power of precarity in conjunction with law can work in favor of a healthy environment. CAE suggests integrating the shared precarity of endangered plant species and endangered social and green spaces in a manner that strengthens and protects both. In many countries, endangered plant species enjoy special legal protection. At the very least, they elicit public sympathy and can function as an ethical hammer for conservationists. Though these plants may be weak as species, they are quite powerful as individual specimens. If that power can be connected with human and nonhuman spaces that are endangered by various capitalist agencies seeking profit and/or power by reappropriating territory held by people unable to defend it, perhaps a sociopolitical symbiosis between plants and people could develop. The plants would expand in number as people plant them for protection, thus addressing the problem of species collapse; in turn, the spaces would have the legal protection accorded to the plants to better resist aggressive or hostile takeover attempts. The kinds of spaces we have in mind are those threatened by unjust development: community gardens, common areas, endangered rural spaces, any kind of squatted territory, and those threatened by extraction industries, including farmland, wilderness, or even suburban aquifers. We plan to place plants at risk and spaces at risk into alliances of precarity that benefit both.

In the interest of our plan, we looked back over the past four decades in the US, and some notable successes in invoking the ESA jumped to our attention. In 1990, the threatened northern spotted owl was used to save its natural habitat—old growth forests—from overharvesting. Millions of acres were set aside in the Northwest, but over the past two decades the debate between conservationists and the logging industry has continued. While the logging industry has managed to continuously reclaim previously protected acres through court actions, lobbying, and even ecological debate (the industry claims that the extinction of the spotted owl is due to the invasion of the competing barred owl, rather than a loss of habitat), the area has been largely saved.

In 2006, conflict between conservationists and the extraction industries heated up over development of Great Plains grasslands in Nebraska and South Dakota. The black-footed ferret, thought to be extinct until rediscovered in 1982, became a symbol of the rich diversity of life on the Great Plains under threat not just from big extraction and ranching, but also from broken public management systems. The introduction of more ferrets helped to extend territories protected from developments and hunters.

The fairy shrimp was used to limit urban sprawl in Riverside, California. Not as adorable as the spotted owl or the black-footed ferret, this species almost had its wetland habitat completely destroyed by developers who covet the flatland on which its rare vernal pools exist. The developers got the better of this fight, but in their revised plan some of the habitat was saved, as opposed to the original plan in which it was all to be completely eliminated. When it comes down to practice on the ground, this outcome seems to be fairly typical. Projects can be slowed or limited, but rarely are they completely stopped.

Consider one of the first classic cases in the US: the snail darter fish and the development of the Tellico Dam in Tennessee. Biologist David Etnier discovered this rare fish in 1973, allowing environmentalists to bring a lawsuit against the Tennessee Valley Authority (TVA) on the grounds that the dam destroyed the limited habitat of an endangered species. Activists used the ESA in filing this suit, a follow-up to one that invoked the National Environmental Policy Act. The Supreme Court eventually heard the case in 1978, and ruled in favor of the environmentalists. Tellico supporters went into action in Washington, DC, to get Congress to exempt the dam from all law pertaining to environmental protection. Through a lobbying initiative, an exemption was placed as a rider to a water and energy appropriations bill. Later, amendment processes were proposed to limit the ESA's power. The dam was completed in 1979.

In the majority of cases, environmental complaints have only thrown a wrench into the works and slowed the process of destruction. In practice, therefore, we must face the reality that this model is imperfect, but it does allow activists to bring some form of power to the table. The TVA still proudly proclaims that no environmental action has ever stopped any of their projects. However, pressure groups representing endangered species or habitats have forced them into negotiations that have impacted the formation of plans. In the case of the snail darter fish, while the land was lost, plans were developed to move as many of the fish as possible to equally hospitable waters. In 1984, the fish was reclassified from endangered to threatened.

We must ask: If charismatic endangered animals tend to have the greatest capacity to motivate environmentalists, why not bring endangered animal species into compatible, threatened environments? The primary answer is because they are mobile. They do not stay where they are put, and they can be hard to find after release, especially if they are a small population. In rural areas in the US, there has to be proof that the endangered species is in immediate danger due to a specific action by a specific agent for the law to be enforced. That means the organism must be close to the disruption. As CAE said above, injunctions are rarely granted, but polluting or destructive activities can be moved off specific sites. Only plants will deliver the long-term presence necessary for this possibility. In urban areas, plants are really the only choice, since the environment is not hospitable to animals. Trees could perhaps also be used in this situation.

Let's take the worst-case scenario in which industry money and political influence are simply overwhelming in a given situation, or legislation has been written with great care to favor industry. We have already seen examples of the former. As for the latter, in Italy for example, environmental protection legislation distinguishes between wild and cultivated plants with regard to endangered species. Cultivated plants, no matter how rare, enjoy no protection under the law. Consequently, CAE's plan would be difficult to enact. The first challenge would be how to classify what is wild and what is cultivated, particularly if it is in a fallow space. While a rare orchid in a hothouse may be easily classified as cultivated, classification might not be so easy in a vacant lot.

Let us assume that we fail to prove our case for protection under the law, forcing us into a second line of defense; could we make a successful media campaign from such an action? To be sure, profiteers have no shame or guilt, but they usually have a public image they wish to protect. If the right species is selected, public imagination can be captured and directed toward a protective position. Increasing the number of stakeholders is always a necessity in bringing a developer to the bargaining table. Obviously, this would require the right choice of species to transplant into the endangered habitat. Choosing a species to save in accordance with its value in the environment is not an option. So many grasses and trees fundamental to ecosystems need saving, but if we are to construct a new coalition we will have to use a more traditional category of choice: aesthetic value. This aspect of the process is unfortunate, but abundantly clear; if the plant is not aesthetically equivalent to the noble beasts at the top of the food chain (awe-inspiring raptors or cuddly mammals), it will probably go the way of the vast unrepresented populations of life forms heading for extinction due to some repulsive characteristic or underwhelming visage. For this job only one choice is reasonable: wildflowers. The image of bulldozers running over big fields of flowers—and not just any flowers, but endangered ones—can have a dramatic mobilizing effect on the one hand, and create a powerful public relations problem on the other.

CAE should also add that this tactic in and of itself is probably not powerful enough to have a large-scale impact. This method is best used when mapped onto other networks of resistance. When public demand for the preservation of natural or managed reserve habitat, or resistance based in stopping the poisoning of an environment by toxic waste are already mobilized, this tactic can function as an extremely useful supplementary activity.

A new alliance between all that is precarious in the human and plant kingdoms appears to CAE to be a highly functional system capable of producing political might; however, questions about logistics may remain. After all, such actions will require many plants, and this need requires a second new alliance among people. One of the divisions that traditionally has been counterproductive for ecological activism is the separation of urban and rural environmental action. While empathy between the two sectors is undoubtedly abundant, bringing the groups together in a practical sense is difficult. This is due not to a lack of desire, but to a lack of infrastructure, which restricts people to acting locally in real space. As much as virtual space helps in organizing and fundraising, ultimately, real-space solutions are needed for real-space problems. For the plant-human alliance to occur, another also has to occur between rural and urban activists and concerned citizens. The new food movement has demonstrated that microfarms and urbanites can come together for a new type of micromarket based on direct local sales. Why couldn't a similar structure be made for a new production of environmental politics, in which a rural contingent can oversee the production of plants while an urban contingent raises money and organizes the distribution actions and media campaigns when necessary?

In Italy

As mentioned above, in October of 2011, CAE traveled to Turin, Italy, to do a workshop on "new alliances" in collaboration with Parco Arte Vivente (PAV). As its name implies, PAV is committed to cultural action that engages ecological commitment as an act of resilience. From its programming, to its architecture, to the remediated site on which it is located, PAV signals a change in the relation of humans to the environment and stands in stark contrast to the modern sites and structures that surround it. Needless to say, CAE could not have asked for a better partner in getting this project off the ground.

We initially conceived of the workshop as consisting of four key parts that would lay the groundwork for the action: CAE would describe the plan and variations for the makeup of the new alliances. We would then proceed by having an agronomist with expertise in environmental law (Daniele Fazio) speak about laws regarding endangered species at the national, provincial, and municipal levels. Next, a botanist/gardener (Filippo Alossa) would talk about local endangered plants and demonstrate how to grow them. The workshop would end with a series of scouting missions to determine the best sites for planting.

When CAE arrived in Turin, Orietta Brombin, the director of education and training activities at PAV, who was also functioning as producer for this workshop, had assembled an amazing team of participants, including our much-needed lawyer and agronomist. Prior to our arrival, the team had already scouted the locations, so we only needed to proceed with the first three parts. The legal session was somewhat disappointing when we learned that Italy made a cultivated/wild distinction when it came to endangered species—cultivated plants were not protected, no matter how rare. However, this would not slow a media campaign, and there did appear to be some gray area as to how cultivation could be proven when discovering a plant in a fallow field. Growing the plants, and acquiring the means necessary to grow a lot of plants, was all quite possible. We were able to choose a flower easily enough (Cupid's Dart, *Catananche caerulea*), although this process was determined largely by the market for commercially available seeds. Obviously, not just anyone can harvest the seeds of wild endangered plants. Fortunately, a variety of endangered plant seeds are commercially available. Finally, we came to realize that the project would have to be extended, as Alossa strongly recommended that we use a natural cycle for growing in natural elements and avoid the artificiality of the greenhouse, as greenhouse-grown plants would be too weak for natural conditions. He suggested beginning to sow the plants in late spring (2012), and transplanting them in late summer (2013). The process is now complete, and guerrilla gardeners in and around Turin are hard at work introducing the plant to endangered urban spaces.

Reversals

In *Dialectic of Enlightenment* (1944), Max Horkheimer and Theodor W. Adorno bemoan the political and cultural regression of reason into a device that furthers a variety of authoritarian tendencies—and conversely, the subsequent failure of the Enlightenment to fully implement its lofty goals of liberty, equality, and progress. As reason regresses, these goals begin to invert and eventually become their opposites. The Enlightenment promised liberation from scarcity, yet no matter how much is produced, the condition of lack continues to grow. The science and technology that seemed the path to perpetual peace and progress instead created a war machine with destructive capabilities that threaten the existence of human life itself; and the rationality that would organize human life for equal benefit of all transitioned to a means to create maximally efficient death camps for the extermination of the Other. Our relationship to nature becomes perverted as nature, too, is perceived as an Other equally deserving of full exploitation and elimination. Cooperative or symbiotic relationships with the earth are eliminated in favor of those of dominion.

This concept is echoed by Ivan Illich's notion of "specific diseconomy." Illich believed that institutions in the capitalist world tended to reverse themselves over time, as they continuously absorb the corruption inherent in capitalism. For example, the idea of public education has tremendous appeal, and free schools and universities as the material manifestation of this ideal would seem like boons to democracy and industry alike as they prepare people for critical, self-reflective, and creative lives. Yet these institutions end up functioning in the opposite manner. They become spaces where mindless ideology is replicated and creativity is undervalued if not completely discouraged. As students are prepared to serve industry as bureaucrats and technocrats, they are taught to tolerate long hours of boredom sitting at desks, staring into monitors, and memorizing the tropes of the free market. Rather than honing their intelligence, students are made increasingly less knowledgeable and are guided by the reduction of diverse knowledge systems to the singular category of training.

Perhaps the most recent incarnation of this notion is Jean Baudrillard's principle of "immanent reversal." The central characteristic of this position is the shift from the dominance of the material order to that of virtuality. Power, pleasure, and seduction are no longer located in the material world. Rather, the house of mirrors that is the virtual world, in which meaning is no longer tethered by referents, becomes the focus of life itself. In this technosphere, the highest degree of illusion produces the highest degree of value and garners the highest degree of praise.

While all of these notions tend to have a negative trajectory, they also point to the possibility that tremendous change is possible. Radical shifts can occur, and if indeed we are at the bottom of the free-market barrel, why can't the next reversal be for the good? A micro war machine of human, rifle, and armored personnel carrier could just as easily be human, brick, and barricade. CAE sees no reason why, through committed struggle, we cannot make ourselves intelligent and creative again, and reverse precarity into a positive, productive power contrary to the present general condition.

Sumatran Tiger

6

Strategy: Lessons in Territorialization

Critical Art Ensemble (CAE) first met members of the art and environmental organization Ala Plástica (AP)[1] while doing a project in Hamburg in 2008. During this time, we were working on an island in the middle of the Elbe River in the neighborhood of Wilhelmsburg—a neighborhood known for housing the working poor, a wide variety of immigrants, and the elderly poor. The commercial infrastructure was less than thriving, but we managed to find a somewhat uninhabited bar, where we would meet in the evenings to discuss how our projects were progressing and various topics in cultural politics both locally and internationally. On one of these nights while on the topic of cultural complexity, AP was insistent that, if complexity was of interest to us, we needed to come to the Río de la Plata watershed region in Argentina. CAE knew we should immediately accept this offer, but logistics are always a problem. Finally, in 2014 we received an award that allowed us to go, witness, and participate in very mature forms of bioregional self-organization, alternative institutions, and strategic cultural resistance that are not so common in the North.

For the past thirty years, CAE has participated in an immensely positive form of deterritorialization. For us, this practice means a reliance on disappearance and tracelessness. In a country that is over-policed and over-surveilled, avoiding theaters of conflict while establishing temporally and spatially limited cultural possibilities that point to relationships for seeing, thinking, and living counter to the imperatives of neoliberalism (with all its alienated and precarious work, gross efficiency, and instrumentalization of everything) appears to us as the most productive way to conduct ourselves as artists. We need to be able to disappear as fast as we bring these oppositional situations into existence. We have to be flexible and fluid, willing to abandon a line of pursuit if met with forces of recuperation and repression, and be able to move to the next relatively frictionless opportunity, regardless of whether it has any relationship to or bearing on what we had previously been doing. Our method is one of reactive tacticality followed by rapid deterritorialization. This form of erasure also ensures that others may appropriate the territory after us, and easily reinscribe it in a way that suits their needs and interests.

Unfortunately, deterritorialization also has a very dark side—one that is very commonly used by the imperial forces of neoliberalism. Upon arriving in Argentina, we received swift and profound reminders of this tendency. We could start with that globally ubiquitous neoliberal model for habitation for the wealthy: the gated community. These were massive tracts of land surrounded by fences, barbed wire, guard towers, and electric gates. The land was stripped of any trace of history, cultural diversity, or emergent social ecology. It was all gone, and sprouting from this wasteland depleted of all its cultural nutrients, power, and pride were the generic markers of neoliberal victory—McMansions, malls, churches, and gas-guzzling cars. It is hard not to feel somewhat sorry for the inmates of these luxury prisons. At the same time, they volunteer to live in them so that they never have to worry about absurdly exaggerated crime waves and never have to encounter, much less interact with, anyone or anything that does not reinforce life as they wish to know it. In the end, this social withdrawal into a fictional smooth space only leads to a profound poverty of consciousness.

But this is of minor consequence compared to the devastation of the rural landscape resulting from extreme deterritorialization by imperial forces, primarily in the form of multinational agricultural corporations. We were warned beforehand, but no warning could have prepared us for what we witnessed in the areas of countryside that have been converted to industrial farming: hectare upon hectare, all the way to the horizon, in all directions, of genetically modified soy. All markers of specificity had been erased. Even the soil, which should be rich, is dead—fully depleted—and now merely a receptacle for artificial fertilizers sold by the multinationals. In lands that once supported grazing animals and a wide variety of grains, fruits, and vegetables, there is nothing but the monocrop of soy waiting for harvest and export—or for the epidemic that kills it all. This is a true biotic cleansing, and among the worst outcomes of capitalist necropolitics. Nothing is left; ecological diversity is reduced to three fundamental types: the soy, the trucks that transport the soy and soy products, and the massive processing plants (as well-guarded and as impenetrable as the gated communities). This sterilized landscape could be almost anywhere; it could be in the US as easily as India. It is a site of generic production as common to the world of neoliberalism as the McMansion. And of course those who benefit from these arrangements present them as massively progressive—a necessary transformation of the Argentinian economy, so that it might be competitive in the global market. However, given the level of policing (mostly private) at the plants, CAE suspects that many people do not agree with this assessment, and that the only real point of consensus is that country life as it has been known for most of the past century is gone.

But why dwell on the darkness when this watershed is so rich and full of life? It is so big that even a monster like Monsanto cannot consume it all. And scattered all around it, on land and on water, are areas the security state has forgotten, or cannot afford to strictly manage. In these areas, experiments in radical biopolitics aimed at ending the current necropolicy that rules the land can take place unmolested by the keepers of the status quo. New alliances are formed, and minds are liberated from misbegotten consensus. CAE's education on *strategies* of reterritorialization began here. We were introduced to a secure network that allowed people who wanted to create a different reality to imagine what it would be, and then to methodically proceed to develop the means to bring it about. In this place, an ecological revolution that comes from the ground up is believed to be possible.

The methodological difference that we found most difficult to adjust to was the relationship to time. Free space we had seen before, but never accompanied by seemingly unlimited time. Here, once deemed a worthy project to pursue, an experiment in biopolitics could take ten or twenty years to ensure it is done effectively and that all concerned are included in the process. As a group constantly worried about instant backlash from authorities and knowing we are lucky to experiment with any viable alternative for at best a year or two (and more typically weeks) before it is either stopped by the law or recuperated, we were stunned to learn that time was a common currency to be used by all, instead of a luxury rarely seen. When time and space in this social configuration meets a bottomless capacity for autonomous action and network construction, anything seems possible. Unlike in the US, where a collective or coalition can only hope that a social or media current will emerge from an action to conjure what Félix Guattari calls a molecular revolution, in Argentina, molecular transformation is an intentional process that works across intentional networks. Here, there is time for networked coalitions to reterritorialize undefended spaces as sustainable ecological systems integrated with expressions of desire.

Most of our experience in the watershed consisted of interacting with experiments in biopolitics, self-governance, and reterritorialization. One of the more historically compelling sites was the cultural center Biblioteca Vigil in Rosario. As the name implies, its original incarnation was primarily as a local library, but it grew much larger into a popular institution of higher learning. During the time of the dictatorship (1976–83), emergent institutions of higher education that also signaled a promise of local autonomy were placed under erasure, and anyone who functioned in any type of leadership capacity was placed under arrest. Books were burned, classrooms destroyed. The library was remade into a disciplinary apparatus, complete with a dungeon in the lower levels. When the dictatorship collapsed, the complex stood as an empty reminder of what never should have been. But recently, local citizens, artists, teachers, and professionals have taken the space back, and a process of reterritorialization has begun. The library has been resurrected, and a push is on to return it to its former glory. There is a daycare center/elementary school, a people's theater (where all productions are free), and plans for a dance theater and music hall. While the material transformation may be slow, the complex again signals that local activism can create transformation in the public interest.

CAE should further note that this process is not a deterritorialization like the political right consistently attempts and all too often accomplishes with its various appropriations. No one involved with the project has any plan to erase the history of the dictatorship in general, nor its manifestation in the architecture. The former deterritorialization by the right is kept visible, and openly discussed in order to create an extra fortification against a repeat of such a process. Those who seek transformation refuse the relationship between history and forgetfulness. This refusal offers a space for necropolitics to enter the discourse of resistance and change, and not only in terms of the atrocities of the dictatorship, and how they cannot be repeated; rather, these activists are able to acknowledge and discuss how necropolitics is a part of the most mundane levels of the bureaucracy and is included in the most common aspects of everyday life. The crucial lesson here is that they understand that the production of life necessarily includes the production of death, and since they are unafraid to surround themselves with this truth, they become capable of pragmatically addressing biopolitics and necropolitics, as opposed to leaving them to the whims of the powerful and the mechanistic outcomes of institutional grind.

Other initiatives appeared to work by way of contrast. Two standout projects were an off-the-grid community on Isla Paulino, and a small, organic farm in the deep rural zones outside Rosario. Isla Paulino is an island vacation community dotted with small homes and lovely gardens. It has campgrounds, beach access, and a small café, and is remote enough that visitors can escape any sense of urban life or the feeling of being surrounded by industry. This is a razor-thin illusion, as the island is surrounded by factories, refineries, and a container port—the true energy hogs of any economy. The island is an oasis in a landscape of ugliness spewing toxicity onto an unflinching earth. Visitors hop on the water taxi, and within a few minutes life appears to be very different. What makes this contrast so great is that all the electricity on the island is created from solar energy. The carbon footprint of this resort has to be near zero. But what makes this community even more surprising is that it was not filled with green activists or folks deeply invested in critical ecological invesitigation; it is populated by people who have come to see solar energy as a part of everyday life on the island. How this transformation happened is unknown to us, but we do know that AP spent a great deal of time there. Though it would be nice to know the process in more detail, CAE is quite happy to see that this did happen, and it stands as a viable example of how holidays could be ecologically considered without surrendering the pleasure of the moment.

The final project that CAE wants to discuss, the experimental farm, is a model for strategic action that drills into the core of economic and ecological struggle. One of the great difficulties in environmental struggle is getting the people most affected by environmental devastion to prioritize it in their daily struggle. Problems of economic and social inequality and education tend to be at the top of the list and for obvious reasons. Those who find themselves in the most precarious of positions only have time for immediate needs. Environmentalism is future-oriented rather than present-oriented for those who must consider food, water, and shelter every day, because daily survival is more than a full-time job. Environmental concerns are a luxury for billions of people. The experimental farm manages to tie all of these issues together, so that working on one is working on all of them. It is a marriage of ecological, economic, and social revolutions. What is being attempted is to avoid the chaos that follows political revolution. Yes, Perónism could assert itself once again and remove the multinationals from the land, but then what? What would land redistribution look like? How would environmental remediation happen? What kind of investment does it take start a viable microfarm? This project's collaborators are working to solve these questions in advance of direct action, and are doing so with great efficiency. Once the model is established, it can be presented to the multitude (and the dispossessed in particular), who can see an immediate answer to their questions of survival. Under these conditions, a combined political and economic popular front could be established that is foundationally green, and that could effectively fight for removal and redistribution in a manner that would benefit a grand majority of citizens and residents. CAE does not want to be overly optimistic, as there are many moving parts in this plan, and it could collapse in any number of areas. However, it is the first strategic revolutionary plan that we have heard in decades that at least sounds viable. The plan lives in the realm of material possibility and not just in the imaginary of utopian fantasy. The following experimental first step in this grand scheme is a material fact.

We were brought to a small ten-hectare farm organized around methods of microproduction that stands as a beacon in the dead zone of genetically modified soy. This initiative brings together farmers, university professors, artists, and volunteers all dedicated to the idea of transforming the landscape into one that is organized via the principles of agroecology as opposed to industrial farming. They hope to demonstrate that small, diverse farms can be profitable without damaging the land or the wildlife and can eliminate the need for large-scale monocropping. If they succeed, it will show that healthy, ecologically based farming is possible and preferable, and that all of the people forced from the land because they could not compete with industrial farming might have the option of going back to it. This initiative is the first step toward major land reform that could repopulate the countryside, taking the strain of the displaced and the dispossessed off of urban centers, and ending an era of the necropolitics of ecocide with which imperial corporations have ravaged both rural and urban environments.

The farm is immensely diverse, and takes full advantage of the many possibilities the land has to offer. Grains, vegetables, herbs, and fruits are all grown on the property, creating a self-sustaining system that does not require herbicides or pesticides. Animals, including pigs, chickens, and rabbits, are also raised for food and fertilizers. The farm represents a complete kitchen. The question, however, is why this small farm will not fail like any other when the goods are taken to market, due to the higher cost of organic microfarming. The answer is that these farmers do not stop at the harvest, but proceed forward by making value-added products. Instead of wheat, they produce flour; instead of on-the-hoof livestock, they produce butchered meat. This lifts them out of the razor-thin profit margins of industrial farming and allows direct-to-consumer marketing, in which they no longer have to acquiese to the corporate demand for macro scalability in order to make a living. *This is a model for earthworks in the twenty-first century* and is a splendid example of artistic and creative vision and action. It is a micro-utopia that is as beautiful as it is practical. We certainly see such microfarming practices in the North, but they are not linked to political and social revolutionary strategies. This project is not the product of drop-out culture (ignoring industrial food economy and developing alternative bioregional markets), but a product of full-spectrum revolutionary engagement.

While CAE came home only more convinced that global capital is an imperial scourge that functions only in the interest of the present generation of plutocratic elites, we also returned to the US with a sense of optimism and a belief that, under the right conditions, patient bioregional strategies can be effective, and should be emulated whenever and wherever the possibility presents itself. Moreover, we found that the necropolitics of ordinary life can be acknowledged, discussed, and to a degree managed, so that as death is produced it is done so in a manner absent of the economic and social prejudices promoted by capitalism.

Note

1. Ala Plástica (founded 1991) describes itself as an "art and environmental organization based in Río de la Plata, Argentina, that works on the rhizomatic linking of ecological, social, and artistic methodology, combining direct interventions and precisely defined concepts to a parallel universe without giving up the symbolic potential of art."

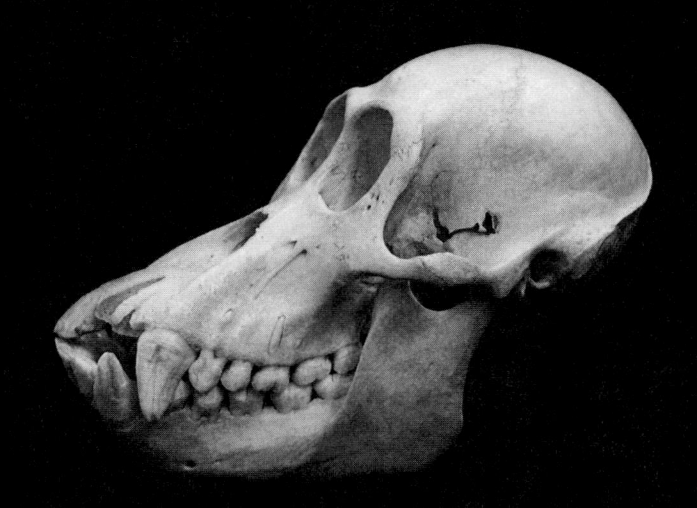

Orangutan

7

The Takeaway

CAE generally considers our books to be works of art—textual performances—as much as we consider them critical reflections. We try to ignore academic conventions and fashions, look for orphaned issues, and attempt to illustrate our ideas and opinions with unexpected sources that rarely harmonize. We aim to keep our texts moving fast, often at the expense of the full explanation that a topic might deserve. Yet, however trim a book might be, it can always be reduced to its basics—in this case, to twelve theses.

1.

Complexity is disruptive to the production of conservationist environmental policy.

2.

As complexity increases, predictability decreases, because all causal variables are not known, let alone accounted for. Because ecological systems are among the most complex and varied systems in the known universe, hypotheses become educated guesses, and the struggle begins simply to establish some modest baseline knowledge through trial and error. Under these circumstances, it is impossible to know all of the consequences of an activated hypothesis in the laboratory of an ecosystem. Science does not know if it has helped more than it has harmed until after the event, and even then only within the great specificity of its objective. Environmental sciences—like art and politics—are full of gambles.

3.

Environmentalism tends to be an aesthetic and/or economic discourse in terms of its objectives, rather than a scientific one.

4.

Necropolitics is a necessary part of environmental discourse and action, as there can be no life without death. Currently, it is explicitly absent, but it is always hauntingly present either as a subtext or as an undercurrent.

5.

Necropolitics, at present, overwhelmingly tends not to be a part of democratic process in Western society (which is not always a bad thing). Even so, we may still be able to produce forms that reduce the extreme economic and social prejudices inherent in capitalism.

6.

Evolution and environmentalism are incompatible because the former makes the latter absurd.

7.

In the face of evolution, humanism and anthropocentrism make pragmatic environmentalism sensible.

8.

Asking "What is best for humans?" in regard to the environment typically yields more environmentally friendly answers than does asking "What is best for the individual?"

9.

Individualism and posthumanism are forms of antihumanism. Antihumanism is a dangerous philosophical position for humans and the environment.

10.

Rationalized environmentalist policy creation and democracy are probably incompatible processes.

11.

Humans are rarely rational actors when it comes to the environment.

12.

Art is potentially an excellent device for communication at the nonrational level in a manner beyond propaganda, and can thus be of great service in the primarily nonrational conversations about the environment. In addition, art can create proof-of-concept models for postnatural assemblages that can serve conservationist efforts.

Dodo

Appendix I

Human Sacrifice and Its Quantitative Representation Or Necropolitics in Everyday Life*

Human sacrifice** is typically assumed to be a "primitive" institution, one that long ago vanished from Western civilization. Unfortunately, quite the opposite is true. The institution of sacrifice lives on. Although much of it is hidden from view in unexpected forms, it remains an essential part of modern everyday life, politics, and economy. A number of ancient cultures, including the Egyptians, the Aztecs, and various Hindu sects, learned to incorporate sacrifice into social life as a visible institution. The practice was legitimized through an association with religious or mystical necessity.

---

*A version of this essay was published as a pamphlet to accompany the exhibition *Acceptable Losses* at the Pacific Northwest College of Art.

So as not to have to redundantly qualify every statement, CAE does not intend this analysis/speculation to be applied to situation(s) in the developing world. Examples (contingent manifestations of sacrifice) offered in this essay may only be applicable to US culture, and not to other complex surplus economies.

** The word "sacrifice" in this essay refers exclusively to human sacrifice.

Through sacrifice, the gods could be appeased, or even bribed, to perform actions beyond the control of either the collective or individual agent involved in the ritual killing. Sacrifice brought together in a concrete manner the worlds of the visible (sensual) and the invisible (spiritual). Anthropologists have speculated that the psychological benefit of this grim ritual lay in its power to relieve anxiety among participants by giving them a sense of control over the uncontrollable elements of existence; an obvious political-economic benefit of ordering death through social ceremony would be enhancement of population management and social control. In cultures where rituals included cannibalism, human flesh may have been a much-needed source of protein. Yet such theories, while they do have some explanatory power, tend to miss the interconnection between the nonrational economy of death and the rational economy of surplus and waste. The willingness to ignore such a connection is one reason why sacrifice continues, unnoticed and incessant, as a standard institution in all cultures of advanced surplus economy.

Our Western propensity for repressing the disturbing aspects of existence means that we are not likely to have a visible institution of sacrifice; at any rate, the legitimizing spectacle that religion would otherwise provide for the practice has melted away under the heated process of rationalization. However, the social functions that human sacrifice once provided must still be fulfilled. Neoliberal society, never content to discard any social action that can either generate profit or maintain social order, allows sacrifice to continue at the margins of visibility. Rather than eliminate the institution, society has driven sacrifice into the under-economy of taboo social relationships and bad objects that should never be brought to mind, viewed, or even named. This realm is the foundation on which the capitalist empire of excess is built.

The under-economy is organized around two kinds of sacrifice, both of which have specific material and hyperreal effects in the over-economy. One is guided by the principle of excess, the other by the principle of autonomy. Sacrifice under the sign of excess is connected to two key economic processes: the production of more than is needed on one hand, and the consumption of more than is needed on the other. To achieve this state of excessive overproduction/overconsumption, considerable numbers of citizens and noncitizens alike must be maimed and killed. For example, consider the use of gasoline vehicles, which most regard as an indispensable right. In light of this context, a minority political contingent claimed that the sacrifice of lives during the Gulf Wars was necessary to provide the Western war machine with a secure supply of fuel, and to ensure that first-world citizens could fuel their cars and heat their homes at a reasonable cost. Though this explanation is widely understood in some sense, it remains a marginal opinion. Our social arena demands that political-economic sacrifice be left unmentioned. The Gulf Wars and their sacrifices were officially sanctioned for the purpose of "liberating" Kuwait and Iraq, and to stop a dictator with militant delusions of grandeur. The morality was visible, but the economic imperative was hidden underneath it, and only briefly became visible through the mediating signs of leftist defiance. While the wars drew some attention to the under-economy sacrifices needed to maintain an excess supply of oil, little attention was paid to the deaths of the 30,000 to 40,000 people who are sacrificed each year in fatal auto accidents. This number is acceptable to most of us in exchange for the freedom to drive—so long as the sacrifice remains hidden and abstract.

Such statistics point toward the second variety of sacrifice, that which is guided by the principle of autonomy. This type of sacrifice, especially when visible, is evidently abhorrent to all political positions except the radical left and true libertarians. For those who occupy this lonely political position, sacrifice is an unfortunate but necessary consequence of the liberation of desire, a compromise that must be accepted as part of the responsibilities of autonomy. For the greater the autonomy given individuals, the greater the sacrifice required. Death and autonomy (that is, the expression of desire) are inherently linked. For example, if we desire rapid transportation at speeds beyond what flesh and bone evolved to survive, we must expect accidents that end in death as well as the possibility that the technology could purposefully be used to kill. Desire can take any emotional form, and it is difficult to accurately predict how it will manifest in action. A possibility always exists that the action will be violent, and hence actively connected with mortality. There is a high degree of emergent uncertainty associated with nonrational activity, and this tends to produce great anxiety; when reminders of our own mortality begin to surface, and the economy of sacrifice becomes more visible, hysteria and panic are typically not far behind. The alternative to facing up to this form of sacrifice and the discomfort of uncertainty has traditionally been the surrender of individual sovereignty to the state apparatus, which is entrusted to legislate what forms of social action will be acceptable. The greater the fear of this form of sacrifice, the more homogenous and repressed the social action required to allay the fear.

Automatic Garage Doors

Every commodity has a degree of risk attached to it, and the possibility for loss of life always exists. Most people manage to keep the uncertainty of life at a reasonable distance, and thereby save themselves the constant trial of wondering whether it is about to end. Yet some cannot keep mortality out of their minds. One situation that conjures this unfortunate state of consciousness is when one loses an intimate to sacrifice. In this case, the object associated with that sacrifice typically becomes regarded as abject by the individual suffering the loss. Often, aggregates of individuals who project death onto the same object form organizations that attempt to reveal the particular sacrifice signified by the fetish object, as well as attempt to destroy the abject object itself.

Much confusion has arisen recently over the nature of the abject. Given recent literature and art exhibitions on the subject, one would think that the abject is defined only by the bourgeois aesthetic of repulsion toward the "filth" of homelessness and toward "perverted" sexual activities. Such things are but one tiny aspect of the abject, if they are in the realm of the abject at all. (Extreme sexual practices may well be a means to escape the abject rather than a means of participation in it.) Any object that mediates the affective apprehension of mortality can become a *temporary* manifestation of the abject. The abject is liquid, sliding into existence at one moment, only to evaporate into nothingness the next. Abject objects are everywhere: they may be safety pins, telephone cords, or automatic garage doors.

Consider the following strange, but true, story: A child is accidentally hit in the head with a lawn dart, is seriously injured, and eventually dies. What follows? An alarm is sounded announcing the need to ban lawn darts (now in a state of limited fetishization). The Consumer Product Safety Commission is lobbied for a law to ban the offending objects. The arguments are simple: "If banning lawn darts saves one life, just ONE, it will be worth it," and "Lawn darts are killing our children!" The manufacture of lawn darts is discontinued. The commission sends out a press release asking people in possession of lawn darts to destroy them, even though lawn darts sold before the ban remain legal and can even be legally repaired.

Once an object is claimed to be abject by a credible organization, its role in the over-economy is assessed. If the object is deemed profitable, and much beloved, or if it provides efficiency in everyday life, then its connection to sacrifice will once again be repressed, and the object will retain its place in the pantheon of either luxury or convenience. (Lots of lobbying, spectacular actions, and other tactics of influence will be used to either destroy or save the contested object's image. Whichever occurs, the perception that triumphs in the legislation process is primarily a product of hyperreality.)

If the object's abject status cannot be spectacularly sustained at a social level, then containment strategies are often used. For instance, many people drown in swimming pools each year, and yet swimming pools (or even better, bodies of water) are not banned. Rather, they are contained. Laws are passed requiring locked fences around pools. The *fenced* pool does not conjure associations with death—hyperreality has declared that this object is not used as a sacrificial altar. Such is also the case with helmet laws for motorcyclists or seat belt laws for drivers. These laws help us to disassociate motorcycles and cars from the under-economy, and keep them clean and visible in the over-economy. Again, we know that approximately 30,000 to 40,000 people will die in the US this year in motor vehicle mishaps.

Recognition of the car as an abject object is extremely temporary. Much care has been taken by the state and the industry to mediate the temporary abject relationships between subject and auto. Signs of safety abound—traffic laws, safety inspections, safety features, the highway code—allowing the auto to be even further disassociated from death. Even more important, however, is the vague intuition of the fairness surrounding this variety of sacrifice. The victims of this ritual seem to be selected by lot. If one has a spatial connection to cars, one enters the dead pool. The greater one's association with the object, the greater the chance of personal sacrifice. Those who love the mechanical extensions of existence, and use their engines to explore speeds that defy the intentions of the flesh, are those willing to trade their lives for forbidden sensations. Mix this desire with rationalized indulgence in various intoxicants and the probability of death continues to rise, along with the intensity of pleasure. Unfortunately the intensity of the violence that often accompanies this sensual exploration is so great that others not receiving the foretaste of paradise are also swept into the vortex of mortality; yet if one drives or rides in autos, such consequences must be recognized. The secondary victim, rewarded at best only by the freedom to drive, is chosen at random, so again sacrifice lurks under the sign of blind occurrence.

Sociopathic Killers

Sociopathic killers are terrorists devoid of political intentionality. This is a popular perception. Like terrorists, sociopaths tend to bring out the worst in people as well as in governments. Terrorists and killers force people to confront the abject in an unstable situation where the horror of the abject seems to consume all that is visible—revealing the malevolent foundation of hyperrationalized political economy. When this process continues for long enough, panic and hysteria are bound to follow. These nonrational motivating impulses are unacceptable in rational society, and yet so many decisions are made on their behalf. The fear of killers surpasses the fear of terrorists—having a political agenda at least makes the latter somewhat predictable, but sociopaths have no intelligible agenda. They are the very icon of the under-economy. They are a frightening reminder that *anyone* can be a sacrificial victim—none shall be spared. Rational argument means nothing when a killer bursts into visibility. Dying in a car accident is far more probable than being the victim of a killer. Yet the news of a killer on the loose inspires panic, whereas the news of a fatal traffic accident—so long as an intimate is not involved—evokes indifference. When one is faced with a killer, individual autonomy seems to come at too high a price. The idea of passively existing at one moment and then being violently thrown into nonexistence the next makes people want to give up their sovereignty to a protector. The police state offers the illusion of total order, a place where such happenings are seemingly impossible, and yet the opposite is true.

The police state, in fact, dramatically increases the odds of violent death. Unlike the nonrational, and hence unpredictable, sociopath, the police state has instrumental reasons for killing (for example, its own self-perpetuation). Giving it the sovereignty to treat life as it pleases only increases the odds of untimely death for everybody (although for malcontents and marginals, the odds are extraordinarily increased). But the hysterical group, caught up in the panic of crime spree hype, has never been known for cool thinking. Is it any wonder that crime bills are passed on the heels of media-scrutinized deaths, or that contemporary campaign platforms are saturated with "tough on crime" rhetoric? Serial killers, macho gang kids, and armed mad junkies cannot be stopped by more police, by tougher sentencing, and/or by more jails. Those who live in the under-economy (or is it those who fulfill the stereotypes of over-economy hyperreality?) cannot be deterred by the disciplinary apparatus of the over-economy, such as fear of capital punishment; that apparatus only works to repress the desires and deter the actions of those who are already members in good standing of the over-economy itself.

Guns

For much of US history, the gun has been considered a necessary tool of production. Whether it was used for the common defense, to clear the land of its aboriginal inhabitants, as a means to procure food (particularly protein), or as a means to legally collect commodities (such as furs), guns were considered instruments of construction, without which a household was incomplete. Guns were also perceived as revolutionary tools: private ownership of weapons acted as a safeguard against tyranny. This latter notion is somewhat anachronistic, since guns are no longer the locus of military hardware, but many still cling to the idea. The NRA tells us that to be good Americans we must be "forever vigilant," and just in case, we must also be armed. These notions have provided conservatives with a mythology and dream of the US that allows them to champion a cause they rarely do: keeping hysteria at a distance and maintaining liberty. Given the conservative record, in which the answer to social problems is to throw those enveloped in them into jail, isn't it surprising that conservatives do *not* want to outlaw guns and put those who possess them in prison?

Oddly enough, in this case, liberals are the ones who want to throw people in jail. For liberals, guns have become spectacularly abject, the ultimate bad object choice. The hysteria over assault weapons in particular is driving the charge. (The actual probability of being killed by an assault weapon is so low that it hardly merits consideration.) The hype generating the hysteria is based on three developments: first, the sacrifice of ghetto inmates has been spilling into suburban visibility; second, the media continuously replays images of sociopaths or terrorists entering shopping malls, cinemas, suburban elementary schools, post offices, commuter subway cars, and other public places and shooting anyone found there; and third, researchers have discovered that when a gun is fired in a household, the casualties are usually household members. For the most part, shootings do not occur in households. The victim is more likely to be a subject enveloped within a specific variety of predatory environment. However, without the stabilizing myths to which the conservatives subscribe, and which help keep the boundary between the over- and under-economies intact, the possibility seems all too likely that one will join the sacrificial pool of victims exchanged for the freedom to possess a gun. Liberals' perception is that a gun is more likely to be used against them rather than on their behalf (yet CAE has never heard a liberal of the over-economy suggest that the police should not have guns). Consequently, the sacrifices necessary in exchange for freedom seem too disorderly and too visible, and hence the reactive call for repression. Reforms such as reducing magazine capacity or closing background check loopholes may prevent some deaths; however, even with maximum repression (a full ban on all guns with mandatory draconian sentences for possession), the violence within an under-economy straining under the weight of capitalist excess will not be stopped. Sociopaths aside, armed citizens in and of themselves are not the problem; the real problem is armed citizens enveloped in a predatory and hyperrationalized economy. Why is the symptom always attacked, and never the sickness?

Sacrifice in War

Sacrifice has always been understood as a necessary component of war. Typically, the youth of a society are sent into battle as cannon fodder, while the support structure (spectacle) of the war machine bemoans their loss and covers their victimization by granting them the status of patriots or heroes. The connection between the spirit world and sacrifice may be lost, but is replaced here by metaphysical notions of national principles (progress, democracy, free markets, etc.). The lack of any absolute grounding for these "sacred" principles is obfuscated by spectacles of misdirection, illusion, and distraction, from parades and military funerals to monuments and TV specials. At the same time, the rationalized contract—that the sacrifice of *x* amount of people will yield *y* amount of profit, prestige, land, and other sacrificial victims—is well known, but unmentionable. Whether this silence is a means of avoiding the dissonance of moral contradiction or a means of avoiding negative sanctions tends to vary.

The most recent wars in Iraq and Afghanistan have taken a turn from tradition. Since these wars primarily consist of battling insurgent nonstate combatants who are poorly equipped to fight their military juggernaut opposition, combat deaths have been greatly reduced, currently standing at 6,519 over the eleven-year period (2001–12) for which we have robust data. Compared with Vietnam (a very similar war), the death toll has dropped considerably. Unfortunately, war requires blood from all sides, no matter how much technology stands in for the flesh. In spite of all the "we support our troops" spectacle, the profuse thanks for their service, the applause as they walk through airports, and the labeling of every soldier a hero, the disconnected American public fails to recognize the true depth of the sacrifice, because it is too abject to witness and consider, and because it too deeply contradicts the social imaginary of what a soldier is and what war is. No one other than the Veterans Administration seems to acknowledge, let alone worry, about the approximately 72,270 suicides among veterans of all wars during this same eleven-year span of time. More American soldiers have died from suicide in this eleven-year span than in Vietnam, Afghanistan, and Iraq combined. The over-economy will never be able to clean the grim reality of veteran suicide of its abjectivity, any more than the coffins of soldiers killed in action returning to the homeland can be cleansed of their abjectivity. Instead, it will be censored and kept invisible and silent in the realm of the under-economy, because the only real solution would be to stop going to war.

Contact Sports

Not all sacrifices end in death. Some victims need only be maimed to fulfill their sacrificial function. Sports are an excellent example. Some may object that sporting practices exist under a rationalized contract: professionals are well compensated for the damage done to their bodies. Perhaps this class of sacrificial lambs does lie on the altar voluntarily, since prior to their pain they are treated as kings, given a foretaste of paradise, and therefore their fate is not so horrid. But what about all the victims sacrificed to produce this royalty? The quality of sports entertainment demanded by consumers is unquestionably high. Direct participation requires a lifetime of training (although spectacular participation also requires a long indoctrination process), and sometimes even biomodification through mechanical or synthetic means is necessary. Since the question of who will mature to join the athletic elite has no certain answers, large numbers of people must begin the grooming process early on so that the pool of potential talent is large enough to yield the very finest athletes. The leftovers from this process must be wasted. Most escape the grooming process no worse for wear, happy to have participated; some, however, do not fare so well. Among this class of throwaways are the sacrificially maimed. They are of all ages: peewees, middle schoolers, high schoolers, and collegiates parade in a stream of biodestruction. Joints, limbs, bones, ligaments, and more are torn, ripped, and shattered, and brains severely injured. Unlike their professional counterparts, these victims receive no compensation other than the fun they had on the way to the altar.

One clear exception to this lighter form of sacrifice is that of the athletes who play in contact sports such as American football, rugby, ice hockey, and boxing. Repeated head trauma (even when not resulting in a concussion) does lead to detrimental effects on the individual player. This problem has been known on a popular level for a long time, particularly in boxing culture, where the fate of sacrificial victims is lovingly referred to as "punch drunk." Somewhere down the road, usually after around eight to ten years, the aggregate of minor head trauma will catch up with contact sport participants, and when it does, their mental capacity diminishes at an astonishing rate, ending in dementia. This condition is known as chronic traumatic encephalopathy (CTE), which recent studies show does not take much contact to occur and can affect contact sports players at a very young age (participation in sports at a professional level is not required). Moreover, the risk level of developing CTE is extremely high if the trauma occurs repeatedly. Current studies put it at 80 to 98 percent likely. These studies are not conclusive (the sampling is very questionable), but the initial data looks very bad, and even if subsequent studies show the risk to be much lower, the outcome will still be unacceptable. CAE believes it will be more than interesting to see what the remedy will be if contact sports are truly as dangerous as they appear. Our bet is on continued sacrifice (and not just for reasons of profit, but for nonrational reasons as well, such as the perpetuation of cultural heritage).

In this case, maiming can serve a double function. Those who fail to become participant athletes still bring profit to the developers of professional sports in a manner beyond offering themselves as material to the sports manufacturing machine. Since these sacrificial victims (the failed athletes) are not ordinarily killed (although such errors do occasionally happen), they become potential perfect spectators. The sacrificially disabled are deeply interested in their sport of choice, perhaps even nostalgic for it, and because they cannot play they are even more willing to pay to watch it being played. The sports industry not only gets product (athletes) from institutionalized sports, but also has its market developed for it free of charge. The harvesting of so many youths for the purpose of developing a sport that can only be watched is surely a sign of the love and sincere desire for the activity. However, it may be a more profound sign of the American love for an ocular order of passivity.

Statistical Representations of Death (Sacrifice)

Numbers regarding the dead should be a case of simple interpretation: the number solely represents the known aggregate of people who existed in an animated material form on earth at a particular time in the past, and who exist no more. A simple fact, that just *is*—but that is not what happens. Statistical representations require complex forms of interpretation because, like all signs, they relay into other signs, slowly building into narratives and discourses and thus becoming untethered from the referents they supposedly represent. Quantity never stands alone, but bleeds into quality. For example, 6,519 US soldiers have died in the wars in Iraq and Afghanistan. Merely revealing this number puts a host of emotions, desires, politics, and aesthetics into play. Meaning immediately balloons and, depending on the cultural and political context, it can take any form. The intentionality of consciousness makes it a near imperative that one interpret statistics about death beyond their referents. There must be a *willing* of narrative. What would a military person, a peace activist, a neoconservative, or a Taliban fighter read in this number? What knowledge would it offer beyond a reckoning of nonexistence for a set of people at a given time? Could we even come to a conclusion as to whether this number is big or small?

Finding ways to tilt the narrative by selecting the "right" statistic, or set of statistics, for the context in which it is placed and the audience who reads it is the recombinant/creative act of the statistician. This too is what makes a statistic boring, telling, outrageous, absurd, or inexplicable.

Acceptable Losses Part 1

Motor Vehicles
2010
Deaths: 33,687
Attitudinal Status: Acceptable
Remedy: Status quo
Source: Centers for Disease Control & Prevention

Recreational Water
2002
Deaths: 4,174
Attitudinal Status: Acceptable
Remedy: Status quo
Source: Centers for Disease Control & Prevention

Terrorism
2001
Deaths: 2,995
Attitudinal Status: Unacceptable
Remedy: Wage 2 wars; radically narrow 5 Amendments in the Bill of Rights
Source: *Washington Post*

Iraq & Afghanistan Wars
As of October 2012
US Casualties: 50,010
Attitudinal Status: Acceptable
Remedy: Continue military presence
Source: US Department of Defense

Lawn Darts
1987
Deaths: 1
Attitudinal Status: Unacceptable
Remedy: Banned in the US & Canada
Source: Associated Press

Medical Errors
2000
Deaths: 44,000
Attitudinal Status: Acceptable
Remedy: Status quo
Source: US Institute of Medicine

Lack of Health Insurance
2009
Deaths: 44,789
Attitudinal Status: Unacceptable
Remedy: Eventually insure 32 million of the 55 million people without insurance
Source: *American Journal of Public Health*

Elevators & Escalators
2009
Deaths: 30
Attitudinal Status: Acceptable
Remedy: Status quo
Source: Electronic Library of Construction Occupational Safety & Health

Foodborne Illness
2011
Deaths: 3,000
Attitudinal Status: Acceptable
Remedy: Status quo
Source: US Food & Drug Administration

Alcohol*
2006
Deaths: 41,682
Attitudinal Status: Acceptable
Remedy: Status quo
Source: Centers for Disease Control & Prevention
*excluding accidents & homicides

Marijuana
2008
Deaths: No Data
Attitudinal Status: Un/Acceptable
Remedy: 847,863 arrests and 7.5 billion USD spent on law enforcement / Massachusetts becomes 13th state to decriminalize marijuana
Source: American Civil Liberties Union

Firearms
2010
Deaths: 31,672
Attitudinal Status: Un/Acceptable
Remedy: Expand gun legislation / Restrict or repeal gun legislation
Source: Centers for Disease Control & Prevention

Fireworks
2006
Deaths: 11
Attitudinal Status: Acceptable
Remedy: Status quo
Source: US Consumer Product Safety Commission

Fire
2010
Deaths: 2,640
Attitudinal Status: Acceptable
Remedy: Status quo
Source: Centers for Disease Control & Prevention

Railroad Car Fumigants
1989
Deaths: 1
Attitudinal Status: Unacceptable
Remedy: New guidelines for warning signs
Source: Centers for Disease Control & Prevention

Lead Paint Toys
2007
Deaths: 0
Attitudinal Status: Unacceptable
Remedy: Recall nearly one million toys; ban distribution of children's books printed before 1985
Source: Consumer Product Safety Commission

Hot Dogs
2010
Deaths: 13
Attitudinal Status: Unacceptable
Remedy: American Academy of Pediatrics calls for redesign of hot dogs
Source: American Academy of Pediatrics

Electricity
2007
Deaths: 389
Attitudinal Status: Acceptable
Remedy: Status quo
Source: National Safety Council

Dog Bites
2008
Deaths: 23
Attitudinal Status: Unacceptable
Remedy: Proposals for breed-specific bans introduced in 86 municipalities
Source: *Dog Bite Law*

Bee & Wasp Stings
2007
Deaths: 54
Attitudinal Status: Acceptable
Remedy: Status quo
Source: National Safety Council

Workplace
2010
Deaths: 4,690
Attitudinal Status: Acceptable
Remedy: Status quo
Source: US Bureau of Labor Statistics

Construction Sites
2010
Deaths: 774
Attitudinal Status: Acceptable
Remedy: Status quo
Source: US Bureau of Labor Statistics

Hot Tubs
1979
Deaths: 2
Attitudinal Status: Unacceptable
Remedy: New warning labels
Source: *Anchorage Daily News*

Cycling
2010
Deaths: 618
Attitudinal Status: Acceptable
Remedy: Status quo
Source: National Highway Traffic Safety Administration

Pedestrian
2010
Deaths: 4,280
Attitudinal Status: Acceptable
Remedy: Status quo
Source: National Highway Traffic Safety Administration

Animal-Rider/Occupant of Animal-Drawn Vehicle
2006
Deaths: 126
Attitudinal Status: Acceptable
Remedy: Status quo
Source: National Safety Council

Machinery
2007
Deaths: 659
Attitudinal Status: Acceptable
Remedy: Status quo
Source: National Safety Council

Airbags
1993–98
Deaths (children): 61
Attitudinal Status: Unacceptable
Remedy: New warning labels on airbags; new warning labels on child restraint products
Source: National Transportation Safety Board

Infant Deaths
2009
Deaths: 26,412
Attitudinal Status: Unacceptable
Remedy: Recommendations to increase access to healthcare
Source: Centers for Disease Control & Prevention

Baby Slings
2009
Deaths: 3
Attitudinal Status: Unacceptable
Remedy: Consumer Product Safety Commission warning
Source: ABC News

Vending Machines
1988
Deaths: 2
Attitudinal Status: Unacceptable
Remedy: Industry-wide warning label campaign
Source: *Journal of the American Medical Association*

Rollercoasters
2005
Deaths: 4
Attitudinal Status: Acceptable
Remedy: Status quo
Source: *Injury Prevention*

Heart Disease
2010
Deaths: 597,689
Attitudinal Status: Acceptable
Remedy: Cut research funding
Source: Centers for Disease Control & Prevention

Cancer
2010
Deaths: 574,743
Attitudinal Status: Acceptable
Remedy: Cut research funding
Source: Centers for Disease Control & Prevention

Acceptable Losses Part 2

Suicide is the leading cause of death by injury in the United States.
Source: *American Journal of Public Health*, 2012

Suicides
38,357
Source: Centers for Disease Control & Prevention, 2010

Homicides
16,259
Source: Centers for Disease Control & Prevention, 2010

Suicides Age 5–34
10,609
8.4 per 100,000
Source: Centers for Disease Control & Prevention, 2010

Suicides Age 35–64
21,754
17.8 per 100,000
Source: Centers for Disease Control & Prevention, 2010

Suicides Age 65+
5,994
14.9 per 100,000
Source: Centers for Disease Control & Prevention, 2010

Female Suicide Attempts
616,000
Source: Centers for Disease Control & Prevention, 2011

Male Suicide Attempts
442,000
Source: Centers for Disease Control & Prevention, 2011

Female Suicides
8,087
Source: American Foundation for Suicide Prevention, 2010

Male Suicides
30,277
Source: American Foundation for Suicide Prevention, 2010

Native American Suicides
441
17.3 per 100,000
Source: Centers for Disease Control & Prevention, 2010

White Suicides
32,010
16.0 per 100,000
Source: Centers for Disease Control & Prevention, 2010

Asian/Pacific Islander Suicides
1,017
6.3 per 100,000
Source: Centers for Disease Control & Prevention, 2010

Black Suicides
2,091
5.3 per 100,000
Source: Centers for Disease Control & Prevention, 2010

Hispanic Suicides
2,661
5.3 per 100,000
Source: Centers for Disease Control & Prevention, 2010

Acceptable Losses Part 3

One in five suicides in the United States is a veteran.
Source: US Department of Veterans Affairs, 2012

Veteran Suicides, 2001–12
72,270
Source: US Department of Veterans Affairs, 2012

Military Combat Deaths, Afghanistan & Iraq, 2001–12
6,519
Source: US Department of Defense, 2013

Military Combat Deaths, 2012
237
Source: US Department of Defense, 2012

Military Active Duty Suicides, 2012
323
Source: US Department of Defense, 2012

Non-Veteran Suicide Rate, All Ages Female[1]
6.1 per 100,000
Source: *Psychiatric Services*, 2010

Veteran Suicide Rate, All Ages Female
12.2 per 100,000
Source: *Psychiatric Services*, 2010

Non-Veteran Suicide Rate, Age 18–34 Female
4.4 per 100,000
Source: *Psychiatric Services*, 2010

Veteran Suicide Rate, Age 18–34 Female
13.4 per 100,000
Source: *Psychiatric Services*, 2010

Non-Veteran Suicide Rate, All Ages Male
18.8 per 100,000
Source: *American Journal of Public Health*, 2012

Veteran Suicide Rate, All Ages Male
29.6 per 100,000
Source: *American Journal of Public Health*, 2012

Non-Veteran Suicide Rate, Age 17–24 Male
15.9 per 100,000
Source: *American Journal of Public Health*, 2012

Veteran Suicide Rate, Age 17–24 Male
61.0 per 100,000
Source: *American Journal of Public Health*, 2012

Iraq War & Global War on Terrorism Veteran Suicide Rate[2]

2003: 26.8 per 100,000

2014: 47.8 per 100,000

Source: US Department of Veterans Affairs, 2016

Iraq War & Global War on Terrorism Veteran Suicide Rate, Age 18–24 Male

2003: 27.0 per 100,000

2014: 124.0 per 100,000

Source: US Department of Veterans Affairs, 2016

Notes

1. In the *Psychiatric Services* study data, "All Ages Female" refers to females age 18–64 years.

2. Among veterans of Operation Enduring Freedom, Operation Iraqi Freedom, and Operation New Dawn who used VHA services.

Siberian Musk Deer

Appendix II

The Entanglements of an Estranged Internaturalist

As told to Critical Art Ensemble

There was once a time when I believed that my relationships with the nonhuman world were of a simple nature. Admittedly, I was predisposed to such a conclusion, since I was a stranger to the urban landscape and was born two decades before suburbia and its insufferability swallowed any sense of cosmic connectivity. Just enough consensual enchantment remained that I could still entertain sublime visions of the cosmos and fantasies of noble savages. Walden did not seem like a dream from another century, but an attainable experience that had only to be claimed. From these cultural beginnings, I could step into an environment of harmony and beauty that awaited those who only needed the courage to see it. These nebulous origins of a mode of thought explaining my place in the world took deeper root one evening as I paddled my canoe on a wilderness lake. Illuminated by the sunset, the earth and the sky appeared as if they had been prepared for angels to descend. In that moment, I abandoned the world, and became an integral part of an inseparable substantive whole. In that brief passage of time, I had shed every point of social and political identity, and was reduced to the simplicity of being-there. I became the living embodiment of the Vedic mystic's fundamental statement, "I am that." Of course, with this realization the experience ceased to exist, and I sat in my canoe weighted once again with all the freight of practical life. I was not dismayed by this sudden pivot from epiphany to anxiety, and took this unique occurrence as self-evident knowledge of a pantheistic universe. My conversion had come. I was beyond the fetish of the sublime, and had regrounded myself in individuation without individuation. I became an internaturalist.

Unfortunately, this forced me to confront the meaning of my freshly claimed point of being. My engine of interbeing transference was empathy. By summoning it I could become all things animal, vegetable, or mineral, and from the variant perspectives surrender any desire for dominion over individuated Others. As I soon discovered, the affirmation of equality among the world's inhabitants was easy; it was the negation of dominion that proved difficult. To refuse it was to turn my back on humanism. No longer could humans be the center and measure of the universe. Their welfare was no lesser or greater than any other. The logic was indisputable, but unfortunately more than logic was at work. Aesthetic judgment was also asserting itself under the guise of reason. A human may accept a judgment ("humans are the most elevated of life forms") grounded only in the desire for it to be true, and due to habitual enactment or witnessing of this belief, it falls within the realm of the real. However, the rejection of an arbitrary or self-interested judgment that is reinforced by the power of social consensus will be perceived as criminality of the worst order. Disciplinary reaction, or sometimes simply the threat of it, pulls the rebellious back into the humanistic regime where equality reigns for some, and dominion is insistent for others.

Not all internaturalists have a profound experience of melting into the continuum of life, resulting in their conversion. Many find their way from the most common of surroundings—a simple sharing of space that is not fully human or nonhuman. Pets are a good example of nonhuman relationships that people can invest with tremendous amounts of emotion, thus developing an empathic bond that is resituated in the familial or even experienced as a bonding of souls or natures. However, seeing oneself in the mirror of the nonhuman Other has its consequences. The pet has its environment, its sustenance, and, to some degree, its movement imposed upon it by the aesthetic (and, at times, utilitarian) decisions of its companion. Here is where the internaturalist may have his cake and eat it too, by recognizing difference and sameness in a common gesture. Because humans are assumed to be more sophisticated in making aesthetic judgments than nonhumans, and have the ability to enact them, it is thought best for them to do so, as long as considerations draped in benevolence are made for the nonhuman. Let us say that the nonhuman in question is a dog. As a protein-seeking creature implicated in the successful satisfaction of its own needs, it will enjoy the bounty of the slaughterhouse, through which it will fulfill a traditional function of cleaning up scraps, only now mediated through the industrialized production of protein. The participation of this dog in the chain of wholesale slaughter of other members of the barnyard is destined by human aesthetic choice. Such choice would be equally aestheticized were the dog given a vegetarian diet, or itself groomed for slaughter. But why stop there? Nonrational aesthetic choice is awash in violence and death. I participate everyday.

For example, I am very fond of various types of alcohol. Every reasonable person, including myself, knows that alcohol abuse is the basis of a major global health crisis, as many cultures of the world have a preponderance of members who quite literally pour a poison into their bodies that attacks every soft tissue in them. The amount of sickness and death caused is nearly incalculable. This, combined with lost public revenue to manage the resulting illnesses and emergencies, not to mention all the alcohol-related violence, would seem to make the production and distribution of alcohol an act of the sociopathic. However, a public consensus exists that alcohol should be affordable, plentiful, readily available, and acceptable to use. Quite honestly, I don't think I have ever lost a wink of sleep due to this demand. To a degree, the same can be said for pets: we want them. Unfortunately, humans are not the ones to be sacrificed for this desire. It is the surplus of pets that bear this burden. Every year millions of pets are killed to ensure that the supply never runs dry, that choices can be made, and that if bonding fails to occur, the animal can be returned or abandoned. What determines a pet's placement in the sacrificial pool are the aesthetic prejudices of the chooser, never the chosen.

Of course, anything can be rationalized away, whether it is pets or animals bred for food. We may bemoan the millions of potential pets sacrificed or the number of animals killed to meet the protein needs of an expanding human population; however, there is an upside to this circumstance, at least when framed by evolutionary biology. While pets and creatures designated as food suffer as a consequence of having thrown their evolutionary lot in with humans, as species they are all doing well. Dogs, cats, cows, goats, sheep, pigs, chickens, and other domesticated species have spread all over the planet and exist in record numbers because of their designations as pets and food. They are genetic success stories in a time of mass extinction. Yet while we may celebrate the evolutionary apex of domestic creatures, we know that the aesthetic categories imposed on the organic realm such as pet/not pet and food/not food are contributing to this same mass extinction and creating very troubling internatural relations.

Bacteria have always fascinated me. If ever a discipline of internatural studies were to emerge, the study of relations between bacteria and the-rest-of-life would be its cornerstone. Unfortunately, the relation of the-rest-of-life to bacteria is a subject that is lost on the general population. Since the emergence of the fantasy of the disinfected body in the late nineteenth and early twentieth centuries, an ever-growing set of fearmongers have been motivated by profit and political power to create beliefs and visions of bacteria that are misdirected, exaggerated, and warped, and that rest outside of the practices of modern medicine and proper hygiene. They would have the public believe that bacteria exist only to create dysfunctional impurity, and thus all relations with them should be severed. The truth of the matter is that the disinfected body (a germ-free organic body) is not possible. For example, humans have a symbiotic relationship with gut *E. coli*; without it, we would die. Some bacteria simply like to live on us, and an infinite variety live all around us. Humans are bacterial hosts no matter how hard we may try not to be, and the environment is always filled with this most differentiated domain of life on the planet (except perhaps under the strictest of "clean room" conditions). The grand majority of bacteria are not a danger to bodily health under normative conditions. Like all creatures great and small, humans have evolved to live with them. Moreover, bacteria are the foundation of every ecosystem. No ecosystem could sustain itself without decay, and bacteria provide this essential function. Unfortunately, advertisers have kept the public focused on bacterial infectants, so the germ hysteria that began in the Victorian era has never really subsided. Even after the invention of antibiotics, the fear of bacteria persists. To my dismay, the personal experience of pain, nausea, and the uncontrolled eruption of a variety of bodily fluids during illness only confirms the exaggerated warnings of various authorities.

The internatural relations between humans and bacteria should be as simple as humans being food and/or reproductive environmental resources for bacteria, while bacteria do the job of systems degeneration and maintenance. However, due to a will for longevity on the part of conscious, intelligent humans, intervention becomes a desired relation. When bacteria break bodily defenses and are free to feed and reproduce anywhere in our bodies, humans strike back with antibiotics, and now do so with such regularity that our actions are working as an evolutionary accelerator, allowing the organisms to become better adapted to the pirated environment beneath the flesh. Bacteria cannot be defeated, and as human relations to them become increasingly conflicted, the only final cause can be human destruction by way of Bacteria Rex (assuming humans do not kill ourselves first).

A second model does exist, in which humans try to productively assist bacteria to adapt to their human environment. Those with intestinal trouble know that the consumption of particular gut bacteria may help alleviate distressing symptoms—the greater the functionality of bacteria in a given environment, the greater the functionality of the environment itself. Symbiosis, rather than a struggle unto death, will commonly lead to an evolutionary path of mutual adaptability. The unfortunate problem is that adopting this model on a social scale would require humans to put their species ahead of their individual egos.

The question of how harmonious relations between creatures can ever exist when some have evolved beyond the mechanical and some have not is as troubling as it is difficult. I find it self-evident that the very nature of consciousness twists existence into unnecessary moral complexities coupled with mortal anxiety. To be free of these concerns would make nature a mechanical wonder in which interdependence and interrelations create a greater whole. Perhaps I am going too far in placing blame on something as poorly understood as consciousness; perhaps the real culprit is intelligence. As evolution has groped blindly along, it has produced many maladapted creatures doomed to be little more than brief stains on the temporal arc of life—but it has created none so dysfunctional in regard to species longevity than this life form that specializes in intelligence. All humanoids are extinct except for one, and while that one appears to be the most intelligent, it couldn't be more maladapted. In a brief one hundred thousand years, *Homo sapiens* are poised not only to eliminate themselves, but to take the higher end of the food chain down the same path into permanent night. Depending on one's disposition toward humankind and the absurdity of existence, this is either one of nature's most comic or most tragic ongoing events.

How is it possible for a beast to exist that does not consider itself a beast? How can a species exist that can simultaneously be, collectively and individually, both on the inside and the outside of ecological relations? But most amusing of all is that a species capable of producing the conditions of its own elimination has evolved, and in its awareness of this possibility seeks to bring its end into ever-greater probability. The engine that drives this grim yet comical occurrence is intelligence, coupled with the cultural ability to accumulate the knowledge produced by it over time. Perhaps Dostoyevsky was pondering the evolution of intelligence when he stumbled upon the inseparability of idiocy and saintliness. Intelligence is an unhappy accident. Consciousness is disruptive.

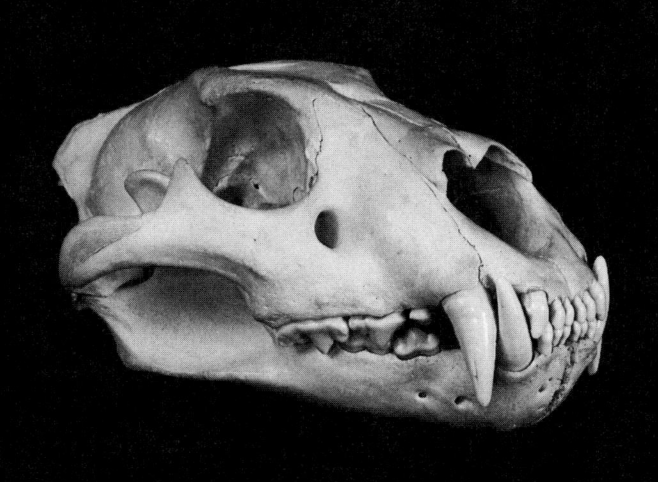

Lion

Appendix III

Anti-systems, Indeterminacy, and Experimental Cultural Practices

At this point in the twenty-first century, few observers of experimental arts would object to the notion that there are currently two distinct and functionally autonomous models. The elder of the two is characterized by expertise in a given specialization that manifests as mastery over a fixed set of materials and advanced technical competence. The task for makers is to radically push or reconfigure aesthetic conventions within the specialization without breaching the specialization itself. As the older of the competing models, its bonds with the institutions of distribution and funding lines are much stronger, so much so that it dominates resources. The junior model (now functioning under many names, including tactical media, hacktivism, experimental geographic practices, culture jamming, artivism, guerrilla art, interventionism, and cultural activism) is characterized by a nomadic tendency to wander through various specializations to acquire and repurpose materials and processes in order to reconfigure culture into alternative forms of perceiving, thinking, and living.

These two models could exist in relative peace (with perhaps a skirmish here and there over common resources) were it not for the insistence of the younger on systemic reorganization of the status quo. In other words, while modest cross-specialization can generally be tolerated (as long as the product appears as "art"), one interrelation that cannot be accepted by the dominant model is the one between culture and politics. These two realms must be maintained as separate at all costs, for failure to do so would reveal the financial apparatus that is the primary driver of cultural institutions, divert attention away from creativity and bohemianism, and potentially erode the system that allows them to profit.

The Call

The idea of what needed to be done, were there to be a sector of cultural experimentalists capable of contributing to resistance against the powers of domination emanating from capitalist political economy, came well before the practice. By the mid-twentieth century, a few key observations and ideas had surfaced. First, the postwar avant-garde as it had been—as a specialization within the specializations of art, literature, theater, and music—had become counterproductive in regard to systemic change. As Roland Barthes famously quipped in *Mythologies*, "What the avant-garde does not tolerate about the bourgeoisie is its language, not its status." The avant-garde is happy to leave the system intact and profit from it as long as it is free to push the possibilities of expression within the system. The system that maintained the luxury market for art happily agreed to this demand as long as product lines remained consistent and distribution remained in its control.

One key idea that enjoyed relative popularity among those who rejected capitalist society was that culture and politics had to be in harmony for systemic changes to occur. Political critique, strategies, and tactics were not enough; there had to be intentional experiments in how to live everyday life with different systems of exchange and participation. The problem at that time was that these spheres of activity remained separate. In 1967, the Situationist call for unification made an appearance: "The critique of culture presents itself as a *unified* critique in that it dominates the whole of culture, its knowledge as well as its poetry, and in that it no longer separates itself from the critique of the social totality. The *unified theoretical critique* goes alone to meet *unified social practice*."

In 1982, this call was echoed by artist collective Group Material in an underappreciated, pivotal work titled *DA ZI BAOS*. One reason this work is so important is that it was unreadable as a specialized product (a subject we will soon return to). This intervention, or perhaps provocation, was installed at the S. Klein building at Union Square in New York City, and consisted of a series of large posters with quotes from mostly local people about their perceptions of culture and social relations. Among the quotes is one from Group Material: "Even though it's easy and fun, we're sick of being the audience. We want to do something, we want to create our culture instead of just buying it." While Group Material would go on to create projects that constituted a *tour de force* in the organization of cultural objects for political ends, they could never escape the confines of art distribution and passive participation. They were not alone, as so many politically active artists struggled with the ongoing contradictions of the avant-garde. While the knowledge concerning the necessity of a new model and thoughts about how this model might look had evolved considerably over two decades, the material conditions to support its manifestation had not.

The Turning Point

In the 1990s, conditions began to change. Notably, the first generation raised with the benefits of the educational reforms won in the 1960s and early '70s had matured and was entering the cultural field. Within these more progressive curricula and models of pedagogy, a sufficient number came to understand the crisis in the production and organization of knowledge. One central problem was that the Enlightenment model of managing the exponential growth of knowledge through ever-increasing specialization within the division of labor was inherently alienating. People were left floating within their hyperspecialized bubbles, unable to connect with other spheres that could advance their area of knowledge or with those who would be consequential recipients. An additional set of intellectual and creative classes needed to be created that could work across disciplines in order to function as bridges between them. By the late 1980s, the first interdisciplinary generation was beginning to establish a beachhead in the universities and the less profitable or prestigious cultural institutions. What these makers brought to the table was a new sense of what experimentation could be. They identified a new box from which they needed to escape: the boundaries of specialization.

Robert Wilson, one of the great avant-gardists of the theater, provides an excellent point of contrast to these newer models of experimentalism in regard to specialization. Wilson states that his practice began and continues with one simple question: "What is it?" (aesthetic indeterminacy). Anyone who has witnessed a Wilson production knows that he does live by this question. Wilson's productions are semiotic riots bursting with wave after relentless wave of unstructured meaning open for endless possibilities of interpretation—and, for Wilson, all the interpretive variations are valid and desirable. He actively invites audience members to collaborate with him by completing the meaning of the visual field (a technique very popular with many avant-gardists). For some, this type of theater can be boring or incomprehensible, or simply not worth the labor, but for those who have developed a taste for co-writing, it's the most satisfying form of art. However, Wilson abandons his question completely in one place: the macro frame of the work is completely stable. Everyone knows they are at a Robert Wilson theater production. The specialization of theater is not challenged, even though its conventions are pushed to breaking points.

In the 1990s, the avant-garde model inverted with the interdisciplinarians—they used common conventions for purposes of readability, but removed the frame. For those who desired to move beyond the limits of specialization in order to interconnect nodes of knowledge and invention, the key signifiers that grounded a given specialization became the point of disruption.

Marcel Duchamp had made the discovery of how to undermine specialized discourse in the second decade of the twentieth century with the invention of readymades and reciprocal readymades. An object could be elevated from the mundane to the privileged by connecting it to the appropriate signifiers that are key to a given specialization. In the case of art, the signifiers included a specific architecture, conventional art object presentation (for example, sculpture should be on a pedestal), and an artist's signature. Even more significant was the theory of the reciprocal readymade, in which a privileged object could be stripped of its key signifiers and thereby reduced to a mundane object (i.e., the use of a Rembrandt painting as an ironing board). Group Material's *DA ZI BAOS* used this reverse method to invert the model of the "readerly" strategy of the avant-garde. While the messages contained within the posters were clearly and reliably readable (conventional), the project itself was unreadable. What is it? A political campaign ad, a billboard, a design project, or just a fragment of the pastiche of wheatpasted trash that litters the walls and fences of every urban center? Within this chaotic anti-frame, with all the key signifiers of "art" removed, art and politics could work together without drawing the usual charges of "impurity," "compromise," or "didacticism" that would make the work easily dismissible within the specialization. This lesson is true not only for art and politics, but also for any other multidisciplinary constellation. The audience can frame such projects in ways that are meaningful to them, and perhaps even more importantly, in ways that the work becomes significant to them. For the interdisciplinary generations, the question "What is art?" is pointless. They have no castle to defend, and are running away from enclosures into open fields.

The Digital Turn

The politicized proponents of interdisciplinary method, resting in a weak network of cultural institutions in the late 1980s and early '90s, did not constitute enough support for a complete split. A technical apparatus was needed that could accelerate the evolution of the model and the network(s). The digital revolution in information and communications technologies (ICT) was a co-development that dovetailed perfectly with the refusal of specialization. In the beginning, this new technical foundation was primarily logistical, having two major consequences. The first, and perhaps most important, was that the new ICT supported the creation of a critical mass of objectors. While finding like-minded people on a local or regional basis could be extremely difficult for a movement in its infancy, having a multi-continental pool of people made networks possible that were impossible before. Through the use of listservs, bulletin boards, websites, and email, ideas could be exchanged at a very healthy rate, and virtual scenes and coalitions were formed. The second factor was that most of this could be done for free or at an acceptable cost. (These traits and activities also partially explain why tactical media, a movement open to all forms of cultural production, primarily appeals to those interested in digital culture.)

This development also changed funding. While no one location had the financial resources for continuous politically charged experimental research, project development and deployment, or peer exchange, many could find a small bit of investment. When networked, new experimentalists could move to where the resources were. Costs could be distributed so that in addition to the established beachhead, there was a nomadic territory in which the movement could grow stronger. Whether a person was working in Bangalore, Budapest, Rotterdam, Barcelona, Beirut, Seattle, or in the middle of nowhere did not matter. There were no more cultural capitals within this sphere of cultural production. This development was liberating in the sense that while traditional cultural capitals and the institutions they contained could still be used and be useful, they were no longer *necessary*. Everywhere was a site of and for cultural intervention. Legitimation through association with geographic territory began to horizontalize.

As ICT continued to rapidly develop in the twenty-first century, the news was mostly good in terms of supporting the autonomy of this second model. (The bad news, of course, was that ICT mapped even more efficiently onto the most predatory and oppressive forms of imperial global capitalism.) Greater access to archives and databases, better tools for organization and mobilization, relative freedom from censorship, and cheaper and more powerful software, hardware, and bandwidth all contributed to freedom from the constraints of traditional limitations. This, in turn, supported independent research and amateur explorations into any field. Alternative voices and those that contrasted with the mainstream could perhaps be drowned out, but they couldn't be shut out, or stopped. Ubiquitous computing begat ubiquitous research, and this allowed the new experimentalists to move into content areas that were once forbidden by specialization (such as science, social science, and engineering), and to speak to and about these disciplines with some authority.

Indeterminacy

As noted, the avant-garde is no stranger to aesthetic indeterminacy. The avant-garde tendency is to search out and explore the extremes of a medium or genre. The extremes along the continuum between overdetermined structure and open-ended randomness have been favored locations for decades. In the case of indeterminacy, the push to interrupt production with random elements—whether mechanical, natural, or social in origin—traveled to a point where aesthetic process was absent of human contribution, and lived as an ongoing process that could be called into existence by anyone at any time. Emblematic of this moment is the work of John Cage, who finally managed to eliminate composers and musicians from music. As he stated, "Music is all around us; if only we had ears . . ." The important point for this essay is that when the disciplinary morph of rejecting specialization occurred in the 1990s, key (un)structuring principles changed as well. Interest in aesthetic indeterminacy shifted to an interest in social and political indeterminacy.

The long-running battle for those coupling culture and politics in more militant ways is explaining the power of these projects to critics and skeptics, while at the same time being involuntarily freighted with the requirement of generating market value. While the geniuses of the avant-garde created tremendous financial and prestige values through singular creative gestures, the experimentalists in the social and political sphere cannot claim to generate much if any political or social capital through our projects. The accusation of failure is ubiquitous, because, seemingly, nothing changes. Of course anyone involved in any type of activism knows that no singular cultural or political activity is going to produce political capital. This form of power can only be accumulated by the many over long periods of time. No one expects that signing a petition or attending a demonstration will result in the instantaneous solution to a major political or social problem. The redistribution of power in its many forms is a long, slow, historical process. This redistribution is not a simple matter of individualized accumulation, as with money and fame, but a long-term collective process in which the whole must be reconfigured.

The question then becomes: How can we organize social relations and territories and arrange semiotic flows in a manner that compels social change? At the time of the bifurcation of the experimental field in the 1990s, suspicion about centralized (and, to a lesser extent, decentralized) platforms was in the air. Concerns about how movements turn into bureaucracies, and how activist arrangements can become oppressive and more reflective of a military order than a liberational one became a preoccupation for those trying to rethink the social relations of resistance. How could the process and aims better align?

To complicate matters further, there were also nonrational considerations that were having an impact on choices made. Where was pleasure in this type of participation? Political activism was and perhaps is a type of service—a sacrifice one makes in order to ensure the rewards that come when the greater good is considered first. A person has to attend the endless meetings, join the picket lines, accept the abuse that follows civil disobedience, join associations and committees, and provide logistical support. The consequence is burnout. Lifelong activists are rare breeds. Those wanting to rethink the social relations of resistance wondered if there was a way to make this category of action more pleasurable and thereby sustainable. Why do we have to replicate the painful social order that we want to escape or even eliminate?

In light of these concerns and questions, it is no wonder that the writings of Félix Guattari began to make so much sense. He knew that the complexity of the social field was too vast to be sorted through the category of quantity. Scale was not a relevant concept—small could be big and big could be small. What was of value was the creation of flowing arrangements or machines that facilitate becoming. Within this field of shifting vectors, outcomes became irrelevant as there is only continuance and flux.

Accumulation and territory were also dispensed with. In this dynamic and distributed social sphere, active participants need only to dart about as particles making new connections where they can in order to increase potential and possibility.

These actions, done without a master plan or final end, took the form of a gamble. Drawing on the avant-garde tradition, going back to Stéphane Mallarmé's roll of the dice, some cultural activists believed that liberational emergent processes could occur even in the face of the failure of a grand multiplicity of single projects in and of themselves. The power of chaos could produce the functional contradiction of a slow revolution. As this had happened on the aesthetic plane with a Dada text-sound piece or a William Burroughs cut-up novel, indeterminacy could function as a tool for an unknown machine that resisted intelligibility in its complexity, but that at the very least could produce leaderless, productive forms of social organization, and at best could change the face of the world.

It was an oddly mystical moment to think that the power of the indeterminate could neutralize so many methodological problems while at the same time being a potential source for public good and social justice. The fear of centralization and in turn bureaucratization as a means to recuperate militant activities began to fade. Deferring to indeterminacy appeared to be a means to eliminate all movement-building mechanisms of the past that had become such a drain on the energy of individuals, and in many cases counterproductive, as with party formation. Now people could follow their own desires as to how they would intervene in culture and politics without having to conform to a master plan. No more meetings, associations, picket lines, or abusive confrontations unless those were the tactics chosen. Acceptance of the indeterminate freed experimentalists to try what may work rather than replicating alienating methods of the past. The new orientation was toward future possibility, which is essential to actual experimentation. In so doing, rigid order and discipline transformed into a gentler system in which rewards were distributed as actions progressed rather than all being directed toward the elusive single reward of final victory. Pleasure could replace sacrifice as individuals could engage social problems and conflicts as holistic entities rather than as resources for a greater cause.

Of course, it's impossible to know if this anti-system worked as well as was hoped. Certainly the political and economic landscape has gotten worse, but we can't know whether the situation would be any better had those involved stuck to more traditional activism. But once this extreme point of distributed organization was reached, a slow reconciliation between various types of resistant forms of organization could begin—that is, between distributed forms (cells, collectives, or affinity groups that share a common perspective but act independently), decentralized forms (coalitions) and resistant centralized forms (commons). Indeterminacy continued to be primarily a feature of distributed networks; however, CAE would like to note that one very radical experiment occurred that brought indeterminacy together with more centralized forms of organization: Occupy Wall Street (OWS). Here, Guattari shook hands with Hardt and Negri.

Occupy's opening gesture was to establish a commons—to show that it could be done, and could be maintained. Once this territory was established, experiments in social relations could begin. How would exchange work? What was the relationship to accumulation? How would the commons regulate itself? How would systems of communication be repurposed and applied? New experiments in biopolitics began to emerge. What were the relationships to sustenance, to temperature, to waste, and even to population regulation? So many of the answers were emergent; discovered in that time and place by the participants. This grand public experiment in social relations was available for all to participate in or to passively watch. The lesson that we don't need politicians, financiers, lawyers, police, or other professionals to govern us was on lengthy display, and ultimately why the movement had to be crushed. OWS vandalized the myth that people are incapable of governing themselves, and once again demonstrated the power of the amateur to find solutions to problems that, in this case, were presented as impossible or too conflicted to solve.

This alone would be a great legacy, but participants were constantly asked (particularly by the old centralist parties, NGOs, and media sources): "What do you want? What is the goal?" There were no programmatic goals, no predetermined outcomes—just continuance. The activity in and of itself was enough. Determination was unequivocally rejected. This disavowal is what separated OWS from Arab Spring and other occupation movements. While the latter had final goals, stopping points, and limits, OWS had none (except those imposed by the police). This is not in any way a criticism of other manifestations worldwide. Occupation movements that topple governments speak for themselves. CAE is only pointing out how historically odd OWS was. In a rather stunning moment of recombinant politics, anarchistic and social democratic principles were cobbled together in a way that has not been seen before or since. Politics without goals had made its visible debut in the over-economy.

Whether it is back in the underground for good is unpredictable, but by sheer example, it put the discourse of inequality into mass visibility, and increased the potential that something might be done about it without putting a limit (which are what demands often amount to) on what that something might be, also indicating a healthy distrust of reform.

Militancy and Recuperation

The bifurcation of experimental cultural production discussed in this essay, including the more anarchistic models and the broader recombinant models of today, has not gone unnoticed by profit-oriented cultural institutions. As one would expect, from the point of view of dominant cultural forms, now that the recombinant cultural option has become autonomous, it needs to be brought back into the fold. This requires that several transformational goals be accomplished. The first is to retrofit the competing model back in with the avant-garde. This is done by claiming that this is not a new model at all, but merely a new material (the social fabric) for artists to use and master in order to bring about fresh aesthetic experiences. This reduction to formalist principles allows for the exploitation of a fully politicized cultural model by stacking the system with "artists" who are willing to decouple from politics and scrub cultural action of all militancy. This decoupling and scrubbing is the second development that must occur. In the final phase, strategies to make salable products out of "social practice" and then to market them are developed. Unfortunately, all of this process is well underway. Beginning with "relational aesthetics" through to the twee disaster that is "social practice," we are seeing a process of recuperation that could end with the actions of the resistant being framed by institutionally friendly brands, dragging us back into the black hole of aesthetics.

Hope and Hopelessness

When describing such bifurcations, an author always runs the risk of presenting a dichotomy stemming from a purity of value that insists that one expression is "good" and the other expression is "bad" in some inherent or transcendental sense. What CAE is trying to offer is a grounded context for the value assertions contained in this essay. In terms of pushing the parameters of expression, we applaud the avant-garde and other associated specialists. Who is not happy that there are Burroughs novels, Richter paintings, Oliveros compositions, or Herzog films? We appreciate them as much as the next art lover. However, if one's focus is the production of culture in order to resist the imperatives of neoliberalism and to develop some alternative to it, then the newly emergent transdisciplinary model is superior in that it has the anarchistic capacity and potential for more contrast, diversity, and independence than ever before (which is not to say that these possibilities will be fully realized). In addition, any optimism about this development also has specific limits. While we are quite amazed that this model exists and continues to evolve at all, that it has some institutional (strategic) support, that it is resistant to elimination via technological means, and that culture and politics can explicitly mix in minoritarian forms, we do not believe that we alone possess the tool that will generate the defeat of global capitalism. This model and its varied applications are a small star cluster in the vast black void of corrupt empire. Sadly, we will not be surrendering our pessimistic sensibility concerning the general condition of global political economy; but we will happily take the small victory that those who stand against the current system have a robust beginning for productive explorations in another area of social relations that we did not have before.